The Conquest of the Microchip

The Conquest of the MICROCHIP

. . . .

Hans Queisser

Translated by
Diane Crawford-Burkhardt

HARVARD UNIVERSITY PRESS
Cambridge, Massachusetts
London, England

Printed in the United States of America
10 9 8 7 6 5 4 3 2

Originally published as *Kristallene Krisen*,
© R. Piper GmbH & Co. KG, Munich, 1985

The preparation of this translation was assisted by a grant from Inter Nationes.

Library of Congress Cataloging-in-Publication Data

Queisser, Hans J., 1931-
[Kristallene Krisen. English.]
The conquest of the microchip / Hans Queisser.
p. cm.
Translation of: Kristallene Krisen.
Bibliography: p.
Includes index
ISBN 0-674-16296-X (alk. paper) (cloth)
ISBN 0-674-16297-8 (paper)
1. Microelectronics—History. I. Title.
TK7874.Q3613 1988 87-24470
621.3817'09—dc19 CIP

Contents

. . . . . .

ILLUSTRATIONS

Following page 110

Foreword to the 1990 Edition

by Robert Noyce

. . . . . .

A technological revolution is not usually recognized until it is over. While it is in progress, the steps are often small and of interest only to those who are intimately involved. The history of the development of technology is one of observing, of asking unanswerable questions, of slowly developing some understandings, leading finally to a headlong rush to find practical applications.

As this book reveals, the history of solid-state electronics is an outstanding example of such a technological revolution. Queisser begins with the early observations of light stored naturally in minerals and the unexpected and unpredictable electrical behavior of crystals. The questions raised by these phenomena could not be answered by science until the twentieth century, when quantum mechanics provided the necessary theoretical structure. Thus are today's marvels dependent on the observations of past investigators and on theories arising from questions asked about seemingly unrelated phenomena.

The continuing story is one of slow advances in the understanding of solids. The structure of a crystal first had to be made visible, which became possible when Roentgen discovered X rays. And a current of electricity could be made to flow reliably through a crystal semiconductor only after researchers learned that such conduction resulted from the addition of impurities, which drastically affect the properties of these solids. Today's electronics depends upon that characteristic of semiconductor materials.

But it is only after predictions are made and useful answers to the "what if?" questions are found that scientific understanding can lead to practical uses. With the first such application, the quest for further understanding intensifies, leading to even more advanced applications.

Those of us who have been involved in the development of this technology recognize that the terms "technological revolution" and "breakthroughs" are used to attract public attention to the progress being made; in reality progress is almost seamless, with pieces of the puzzle continually being put in place until a coherent picture emerges. The history of solid-state electronics can serve as a reminder of this truth. Fortunately, unlike our exhaustible natural resources, knowledge is an expanding resource. As it is used, it generates additional knowledge, which will become the foundation for the "breakthroughs" of tomorrow.

The microelectronics industry has become a cornerstone for the development of industrial countries and thus will have a decisive influence on the life of all societies in the next century. As Queisser describes in the latter part of the book, an intense international competition has arisen between the United States and the Asian countries, especially Japan, for dominance in the field of microchips. Indeed, since the first edition of this book was published, this economic competition has extended further down the "food chain" to include semiconductor manufacturing equipment and materials. An acute awareness of the impact of such developments has led to the formation of a joint Semiconductor Manufacturing Technology initiative (SEMATECH) in the United States. Even the European nations, which until now have remained relatively withdrawn from the fray, have begun to react to these economic and political developments by considering a joint semiconductor research and development venture called JESSI. The future will undoubtedly bring new alignments of nations as well as technological developments that we cannot now foresee, just as no one in Marconi's time could possibly have foreseen our present world of advanced semiconductor technology.

Preface

. . . . . .

THE great epochs of human history—the Stone Age, the Bronze Age, the Iron Age—have all been named for solid materials. Mastering each new material has spurred the great cultural advances of humankind. Now, in the second half of the twentieth century, the microscopic world of the silicon crystal provides the stage for a new epoch, the Silicon Age.

If ever a product has been brought close to perfection, it is the silicon crystal, in which every atom is fixed in a specified position. The electronic pathways within semiconductor chips and transistors depend on the precision of this lattice structure. As the new field of microelectronics has developed, the older materials technologies have become obsolete. In earlier times the technology of a new solid was mastered through experiment after experiment, trial and error, clever adaptations of chance results. But when researchers began working with sensitive semiconducting crystals, they found these time-honored methods insufficient. Only by painstakingly using scientific theory and mathematical analysis could the microchip be conquered and reliable components produced. No wonder, then, that traditional technologies face problems today. The incomprehensibly minute structure of the crystal threatens old occupations and devalues traditional skills, even as it makes the unimaginable real.

As a young solid-state physicist, I had the good fortune to participate in the birth of the microelectronics industry in California. The risks and challenges of that fascinating adventure in high technology have held me in their spell ever since. In this book I describe, without using technical jargon or mathematical formulas, the early history of the new science of crystals as it developed in Europe; then I trace the technological applications of that science as it moved westward to the United States and then to Japan. I also introduce some of the scientists and researchers whose contributions were fundamental to our understanding of the physics of crystals. Only when science had revealed the fundamental properties of atoms could researchers tackle the enormously difficult task of describing how atoms interact within the crystal lattice. After depicting the rapid evolution of this technology and its previously undreamed-of applications, I take a look at the future promised by microelectronics.

The original German edition of this book, published in 1984, aimed to educate the public about this new field of science and business, which has been rather neglected in Europe. The book may have succeeded in stimulating some public awareness and action in West Germany. American readers, however, have long been aware of the significance of semiconductor microelectronics, with daily newspaper headlines to remind them. Therefore the English-language edition, translated by Diane Crawford-Burkhardt, was extensively recast for an American audience by Howard Boyer, Lisa Rosset, and Peg Anderson. I would like to gratefully acknowledge the translator and the editors for their pleasant cooperation.

The Conquest of the Microchip

1

Crystal Crises

. . . .

A PERSON with vision in only one eye tends to misjudge spatial depth. So Marchese Guglielmo Marconi, blind in one eye after a near-fatal accident three years earlier, in 1911, was finding it difficult to set up his experiment. He had succeeded in attaching the thick copper leads to the brass clamps on the little wooden board in front of him. But placing the needle's fine point precisely on the shiny surface of the bluish gray galena crystal—that was not so easy.

Marconi and His Wireless

In the room outside Marconi's office, a young assistant waited impatiently, eager to know the outcome of this scientific demonstration. The little receiver was such a simple construction—a crystal of the mineral galena and a spring with a sharpened needle tip—but it promised to make "radio" a household word. With headphones and a little luck, one would be able to hear radio-transmitted words and music. No cumbersome, expensive equipment would be necessary. Every home would be able to afford this kind of receiver. Marconi Wireless, Limited, as well as the rest of the competition in the industrialized nations, would market transmitters, loudspeakers, and other communication equipment. These high hopes all centered on the small dark crystal.

Marconi himself was skeptical about such commercial possibilities. The father of wireless telegraphy, he felt that his invention should be reserved for the experts and not become a public toy. His success in sending a Morse code message across the English Channel and then across the Atlantic Ocean had made it possible to send messages to ships at sea, to relay distress signals, to provide help quickly. But to use the wireless for banal music and perhaps even lowly advertising—that seemed traitorous to him.

Even as a child, the Bolognese nobleman had been excited about electricity. His parents had hired a private tutor for him, and Marconi would sneak into Professor Righi's physics lectures at the university. By 1895 his first experiments, conducted in the garden of his parents' summer home in Pontecchio, had succeeded in transmitting sound with electric waves. One year later, on the roof of the British Postal Ministry, he proved that it was possible to send information through the air. Still the experts were skeptical. The great Jules-Henri Poincaré, prince of French sciences, had proven beyond all doubt that radio waves would never be able to travel from Europe to America. Because they did not follow the curvature of the earth, they would be lost in space. Marconi decided to try sending a signal across the Atlantic anyway, relying on the "principle of optimum ignorance." And he was successful. Of course, after the experiment had succeeded, the experts outdid themselves in providing theoretical explanations.

Marconi particularly doubted that the crystal receiver would work. Clamped in its little brass pot, the crystal was supposed to receive and convert the electrical energy of radio waves into sound. Marconi had had all sorts of problems with unreliable radio receivers. In transmitting, the use of higher antennas, supporting metal reflectors, and more power had solved some of the problems. But from the beginning, catching the spreading radio waves and converting them into an audible sound had been fraught with nearly insurmountable difficulties. One solution to the receiving problem had resorted to a pretty dubious tool, not at all well understood, that had been found by chance during laboratory experiments. Iron filings placed between two little wires housed in a glass tube would react to incoming electrical waves. Even weak signals changed the electrical resistance between the wires. The learned physicists could not explain this effect; the practical-minded Marconi had simply used the tool, even if it functioned only by hit or miss.

Marconi's difficulties with such unreliable receivers made him feel uncomfortable about having the big company that bore his name take another risk, this time involving a precarious crystal as the core component. In no way was it certain that touching a needle to a crystal's surface would work. One had to patiently search for the spot, not readily discernible, where reception was possible. Some crystals were completely useless as receivers, even though they came from the same source as other, perfectly suitable samples. How could you sell the public that kind of unreliability?

The marchese took off his headphones. The signal reception with his crystal was fairly good. His technical assistants had apparently found an exceptionally useful specimen for the demonstration. Sunk in thought, Marconi paced in his study, its walls covered with documents, awards, pictures, and one framed postage stamp that was his pride and joy.

There was another reason for Marconi's discontent. In 1909 the Swedish Postal Service had designed a stamp in honor of the two recipients of the recently created Nobel Prize for physics. The heads of two men looked out at the observer. On the left, serious and severely elegant, was Marconi, just thirty-five years old. On the right, with a full beard and little rimless glasses, was Ferdinand Braun, the inventor of the crystal receiver. Between them was a radio device with an antenna. Linking the two men was a garland with the words "Nobel Prize" written in Morse code. Braun was Marconi's chief rival in the field of telegraphy. Braun had discovered that galena crystals had the unusual property of reacting to electric waves, and now Marconi's company was considering the use of crystals for the radio market. The two competitors could scarcely have been more dissimilar: the thorough German professor versus the daring Italian dilettante.

At first, Marconi had felt insulted at having to share the prize with Braun and had threatened to reject the award. His friends reminded him that the prize for physics had already been shared twice, in 1902 and 1903. They soothed the marchese and helped him realize that Braun's scientific achievement had also contributed to the introduction of this daring new technology. Journalists who had traveled to Stockholm in the hopes of reporting on a dramatic personal conflict surrounding the awards ceremony had to leave disappointed.

In his Nobel acceptance speech Braun referred many times to the new field of wireless telegraphy, but in passing he also mentioned two of his other discoveries. One was an instrument that was of interest only to a few scientists. He had shown that in a glass tube from which all the air had been pumped out, an electron beam would move back and forth. Because the beam had such small mass, it would change direction quickly when, for example, a small magnet was brought close to the tube. The other discovery was that when a metal needle was placed in contact with a galena crystal, an electrical current flowing between the needle and the crystal would exhibit strange behavior. In 1909 nobody dreamed that fifty years later all of humanity would be greatly influenced by these two "trifling" discoveries. Braun's tube would become the television receiver; and the needle and crystal device would give birth to the microelectronics industry.

Braun's Discoveries

Ferdinand Braun was born in Fulda, Germany, in 1850. As a bright student attending secondary school, he was especially interested in the natural sciences, which his teacher emphasized as an important element of cultural life. Braun learned to open his eyes and critically examine even apparently simple, obvious processes in nature. His teacher was only half surprised when the fifteen-year-old student came to him one day with a complete manuscript on the science of crystals. But even under a pseudonym, it would have been unseemly for a young student to publish a book. Later, though, he published an article entitled "Water." In skillfully conducted experiments, Braun had produced water very pure and free of chemical additives and had cooled it so carefully that it did not freeze until it reached a temperature 12 degrees centigrade below its regular freezing point. Braun was surprised to achieve a condition so far from equilibrium. He realized then, and never forgot, that changes of state, such as the change from a liquid to a solid, can be influenced by even the smallest additions of foreign matter.

Braun received his doctorate in 1872 at the age of twenty-two. As an assistant at Würzburg University, he began to experiment with electricity, which people all over the world had suddenly become interested in. He examined many different materials for their ability to conduct a current. Würzburg University housed a huge collection of minerals, which he made good use of in his systematic measurement of the flow of current through solids. Later, as a teacher at the Thomas-Gymnasium in Leipzig, he was able to use the school's mineral collection and thus continue his research.

The American historian Thomas S. Kuhn has explained how everyday science differs from revolutionary science. Normally scientists patiently gather new information that they can fairly easily incorporate into the existing system or paradigm. The mosaic pieces that they add enlarge a concept upon which the experts basically agree. The generally accepted rules of research determine how new facts are to be acquired and categorized. If a researcher's results differ from the paradigm, he first looks for errors in his own work, approaching his results with the greatest care and reservation. But if a contradiction remains after all the tests have been conducted, it may signal the beginning of a scientific revolution.

On November 14, 1876, the audience attending a lecture given by the Leipzig Physics Society was witnessing a scientific revolution, but most of the learned listeners were unaware of it. Physicist Ferdinand Braun, recently promoted to permanent senior teacher, was lecturing about "Deviations of Ohm's Law in Conducting Solid Bodies." Sitting right in the first row of the auditorium were the university professors, all internationally recognized experts in the field.

Braun reported on his simple, yet carefully conducted experiments on conduction of electricity through pyrite, galena, and other sulfur-containing crystals. With a metal needle point, he directed the electric current into the crystal. The current would flow out through a large surface contact. In this way one could test how strong the current flowing through the crystal became when voltage from a battery was switched on. Georg Simon Ohm, also a schoolteacher, had demonstrated in 1826 that current and voltage are in a fixed ratio to one another; if one doubles the voltage, the current also doubles. The direction in which the current flows plays no role; there is no difference between backward and forward.

Braun explained that Ohm's law did not apply for his crystal. In this case the strength of the current depended on its direction. In one direction, his crystal permitted a strong flow; in the opposite direction almost no current could pass.

The Leipzig Physics Society witnessed a further unexpected event. The resistance of a substance is, by definition, the applied voltage divided by the current. The resistance of all good metallic conductors does not depend on current or voltage, according to Ohm's ironclad law. But in these crystals electric resistance seemed not to be a fixed quantity. Instead, it varied according to the strength of the current. This conclusion, bordering on heresy, was so significant that Braun carefully described and demonstrated every connection, every switch, every device in his experiment. The results could not be denied: electric resistance in the crystal lessened when the current was strengthened.

Braun, a well-trained physicist, knew perfectly well how badly his results fitted the traditional view of electrical conduction in crystals. His findings deviated from two fundamental principles of physics: reversibility and linearity, two principles that have been proven over and over in many of the natural sciences. An elementary process should not depend on its direction; whether a particle moves to the right or left should have no real influence on the nature of the movement. Reversibility of elementary processes is a basic physical law. But Braun's experiments appeared to overthrow the principle of reversibility. Similar asymmetries between left and right would cause comparable commotion in atomic physics eighty years later.

Ohm's law, so fundamental to the science of electricity, was unmistakably flouted by Braun's experiments, which must certainly have shaken up his audience in Leipzig. On that day in 1876 no one could have dreamed of the significance of this discovery. These two violations of reversibility and linearity were to become the essential prerequisites for modern electronics.

What Braun had demonstrated was the principle behind the rectifier. When the metal tip touched the crystal, current could flow in only one direction. With such a crystal, one could generate direct current—that is, current flowing in one direction only—from alternating current, in which the direction of current flow changes continually. Such rectifiers are important to electronics, for they enable current differentials to be maintained.

Natural phenomena tend to seek equilibrium: hot water cools to the temperature of its surroundings, rain and wind eventually level out the differences between mountains and valleys, a drop of ink disperses in a glass of water. Every organization, whether in social life or technology, must continually work against this natural equalizing tendency. Contrasts, differences, states of dissimilarity must first be generated and then maintained. Without a device that forces flow in only one direction, it is impossible to maintain an imbalance. The valve of a bicycle tire allows air to pass in only one direction and build up a higher pressure in the tube than outside. The heart, both a pump and a system of valves, is a one-way system, transporting oxygen-rich blood to needy cells and returning the oxygen-depleted blood to the lungs. Similarly, to allow an electric charge to accumulate at a desired spot and not flow backward, requires some sort of electric valve. Braun discovered that crystals were able to function in this way, but no one knew why.

The electrical resistance of Braun's arrangement of a needle tip on the crystal was not a constant value but varied according to the current itself. This unexpected discovery meant that current had a nonlinear relation to the applied voltage. The nonlinearity of resistance also counteracts uniformity and proportionality: a strong current encounters little resistance and is more readily transmitted, whereas a weak signal encounters high resistance and passes less readily. Such a nonlinear resistance can be used to construct a threshold, accepting current only above a certain minimum strength and suppressing everything below that level. The only differentiation now is between current and noncurrent, between one and zero. This reduction to a simple yes-no decision is the principle behind modern calculating machines. Braun's crystal offered the opportunity of realizing this advance back in 1876, but it was many more decades before anyone could even come close to explaining what Braun was demonstrating with his experiments.

Braun was unable to obtain consistent data and could not generalize his findings. The secret of the crystal seemed to lie in a thin layer just below the surface. And something about the contact, the interface of needle and crystal, was critical. Braun repeatedly wrote of the "peculiarity of the contacts." The unusual findings were especially prominent when one of the two contacts was small: "I therefore mostly used a spring-loaded wire, which I pressed against the crystal." This clever idea of applying a springy wire with a minute point was Braun's inspiration, but only in the next century would its full impact be felt. Only a science of the behavior of electrons in a crystal, "electronics," would supply the explanations Braun was seeking.

From Cathode Rays to Coherers

In 1895 Braun was appointed full professor for physics at Strasbourg University in the imperial state of Elsass-Lothringen (Alsace-Lorraine). The government in Berlin was spending a lot of money to make this university a showcase, and the Strasbourg Institute of Physics became a busy center of research, with excellent equipment and personnel.

Braun continued to be interested in conduction of electricity. Now physicists were pumping air out of glass bulbs and feeding electricity into the tube through a wire to investigate the motion of electrons in a partial vacuum. Electrons could be coaxed out of the white-hot wire and shot through the tube. Finely powdered crystals deposited on the wall of the glass tube would glow when bombarded by these electrons. Braun was drawn to this exciting new field of research, and his involvement brought him once again into contact with crystals.

Braun searched for a device to measure the time lapse between electric oscillations. In those days electricity was just beginning to make its presence known in everyday life, lighting up cities and driving motors. Generators supplied alternating current, which changed polarity rapidly. But the changes over time could not be measured. A dial gauge was too slow; it merely twittered back and forth a bit and could not keep up. The measuring device had to have the smallest mass possible to keep pace with fast variations. On February 15, 1897, Braun demonstrated a new instrument at the Strasbourg Institute. The current whose time structure was to be measured was connected to magnet coils near the walls of a vacuum tube. These coils would deflect a beam of electrons in a pattern that varied as the current varied. Stronger current made the beam deflect more. When the electrons hit the powdered luminescent crystals on the wall of the tube, they painted a glowing curve that represented the variation in the current being measured. "Braun's tube"—the cathode ray tube—has become the oscilloscope. This device for viewing oscillating electricity is one of the most important instruments in electronics and even more widely known as the picture tube of television sets all over the world.

When the tube was demonstrated in Strasbourg, the audience enthusiastically watched the glowing dot dart around the phosphorescent screen. One could finally prove that the Strasbourg Electricity Works really did generate a neat, sinusoidally varying current.

The popular enthusiasm for electricity led to research in wireless telegraphy at the end of the century. Braun's research in this area eventually netted him a Nobel Prize, but it also entangled him in a web of external interests. The great seafaring nations of the time—Great Britain, followed by that upstart, the German Empire—were well aware of the significance of wireless transmissions from ship to ship and ship to shore. Shipping companies and marine insurers showed a keen interest in everybody who knew, or claimed to know, something about electromagnetic waves, antennas, transmitters, and receivers.

In 1890 the French physician and chemist Edouard Branly had invented the first detector of radio waves, a little glass tube filled with iron filings. His discovery greatly accelerated the pace of wireless telegraphy, but nobody, including Branly, could understand exactly what was happening between those iron filings when they were affected by radio waves. To be sure, the resistance of a current passing through the filings lessened when radio waves passed through. This reduction in resistance acted like a switch and could easily be measured by a secondary circuit used for wave detection. But the reason for the effect of the waves upon the filing crystals remained obscure and is still obscure today with more definitive detectors available. To camouflage their embarrassing lack of understanding of this cantankerous device, scholars had agreed on an elegant name: they called Branly's device the "coherer." If you practiced a little magic, you sometimes could get the strange tube to receive high-frequency radio waves. A Morse code signal in a radio wave could thus be detected with a loudspeaker or a ticker. After the current had flowed through, the filings had to be returned to their previous loose state with its high level of resistance. So the technicians tapped the tube to separate the filings. Later, they even added clever automatic tapping devices.

Everybody tried to improve the coherer, but no one understood its scientific principle, so there were no guidelines for promising directions of research. Tinkering was the only approach possible, and it often seemed to yield small improvements. Competition was fierce to fabricate better detectors, which were the weak link in telegraphy. Industrial espionage became notorious because of this enigmatic little gadget, and devious schemes to steal competitors' secret witchcraft were devised.

Science's capitulation to random tinkering was nothing new for Braun. The dilemma of the coherer strongly resembled that of the needle and crystal device from his Leipzig days. Solid bodies, crystals, played a role in both cases of unusual current flow. Moreover, small surfaces of contact seemed to be the key not only to the needle and crystal device but also to the iron microcrystals in the coherer glass. Large filings or solid iron simply would not work.

Once again Braun took up his old Leipzig studies, but this time with a different aim. Unlike his competitors, who were only interested in improving the coherer, Braun wanted to replace it with a new system using a single large crystal and a fine-point contact. He was successful: the crystal receiver became equal to and sometimes better than the iron-filled tube. Although the physical laws governing this new receiver were no better understood, the device was definitely easier to use and looked more scientific.

The telegraphy pioneers correctly sensed that one solid crystal was much more likely to furnish a reliable technical solution than the loose mixture of iron crystallites. And so all telegraphy companies, including Marconi's, pinned their hopes on this new detector.

Vacuum Tubes—A Detour

Among the twists and turns in the path leading to modern electronics was a lengthy detour that was unwittingly initiated in 1883 by Thomas Alva Edison. He was investigating ways to improve his most famous invention, the electric lightbulb. The problem was that its glowing filament steadily dissipated, leaving only an ugly, obscuring deposit on the walls of the bulb. Edison, a practical man but also a scientist, knew that only by precisely measuring the migration of carbon from the filament to the walls could the problem be solved. He therefore introduced a little piece of metal, a "grid," into the lightbulb to measure the current of electrons flowing through it. To his great surprise, Edison saw that he could regulate the current by manipulating the amount of voltage on the metal grid. This "Edison effect" might have enabled its discoverer to add the electron tube to his long list of inventions, but at the time Edison was too busy with power plants. Not until 1904 was the device investigated again, this time by John Ambrose Fleming, who created the first vacuum tube rectifier based on the Edison effect.

This invention dealt the death blow to Braun's and Branly's crystal receivers. The electron tube permitted current to flow in only one direction, just as those older, unreliable crystals had. But the glass tube had the advantage in that one could perceive and manipulate all the individual components. Everything could be seen at a glance; one could forget about the crystal's enigmas. Since the stream of electrons flowed only from the hot filament to the colder plate within the tube, and not in the reverse direction, when a negative pole was attached to the cold plate, the simple electron tube served as a rectifier.

The price for this convenient setup, however, was high. Bulky glass containers, with the air carefully pumped out and supplied with feedthrough leads, were necessary. Moreover, a lot of energy was needed to sustain the red-hot temperatures. Crystals were much smaller, more convenient to produce, and required less energy, but because they were less reliable, they were replaced by the vacuum tube. Eventually scientists recognized the inefficiency of vacuum tubes for most applications, but not before many years of research were spent on this form of macroelectronics.

Around 1920 tube electronics took another step forward when the amplifier was developed. Whenever an electric signal is transmitted, it gradually becomes weaker and less audible and must be refreshed. The components used to amplify a signal must draw only small amounts of power to control much energy supplied by some reservoir. Three accesses are necessary: the input and output of the main current and the control element. The principle of the amplifier was discovered by many people, including two Americans, Irving Langmuir and Lee De Forest. These inventors introduced a third metal element, a control grid, into the glass bulb. As the current flowed through this grid, which acted like an adjustable hurdle; a small change of voltage could effect large-scale changes in the flow of current. A weak signal at the control electrode could now generate a strong change of current flow: amplification became possible. Crystal receivers could not yet fulfill this important function.

De Forest, who held a Ph.D. from the University of Chicago, can be given the credit for introducing classical language into electronics. He named his tube with a built-in amplifier the "audion." Other impressive-sounding Greco-Roman names for vacuum tubes would follow: triode, pentode, klystron, magnetron. With an audion, one could make signals loud and clear; the controlled energy was strong enough to run a true loudspeaker. Headphones were no longer needed to hear a radio signal. The entire family could sit around the radio and listen to their favorite broadcast.

De Forest's amplifier could also generate oscillations and hence be the source of radio waves. The trick, called "feedback," is accomplished when a small oscillation, a change in the strength of the current, is fed back to the control electrode, which amplifies the current. Such powerful oscillating currents can be used for broadcasting. The previous source for oscillatory current was an electric spark discharge, but it was quite unreliable. De Forest's device was much better understood and was based on sound scientific principles.

In the good old American way, De Forest first tried to set up his own company and make money on his invention. But he was not a very good businessman, so in 1911 he gave up the company and went to work at the recently established laboratory of the Federal Telegraph Company. De Forest moved from the hectic East Coast to California, which was still sparsely populated, and settled down in Palo Alto, a sleepy little town set among fruit groves, the home of the university founded by railroad king Leland Stanford.

Because of mass production, even the complicated vacuum tubes could be produced cheaply. By 1929 the United States was manufacturing more than ten million radios a year. In 1930 De Forest was elected president of the new and flourishing Institute of Radio Engineers. In his acceptance speech the "father of radio," or at least of the radio tube, bitterly scolded American radio stations for exploiting his invention with their petty, uncultured broadcasts and primitive advertising.

The storm of new technology, business, and money roared through every physics laboratory during the early decades of this century. Whoever did research in the field of electric waves was suddenly lured out of the staid academic world of the Maxwell theory and Hertz waves and into the hurly-burly of business competition. In 1900 General Electric, the growing giant that had taken over Marconi's American affiliate, established the General Electric Laboratory and appointed Willis Whitney, a doctor of philosophy from Leipzig University, to head it. Other famous scientists and engineers who joined the lab included William D. Coolidge, also the holder of a doctorate from Leipzig, and, of special note, Irving Langmuir, later a Nobel Prize recipient. Having a staff of scientists did not always assure a company of the lead in industry; the scientists' careful, lengthy deliberations and reservations often were a crippling disadvantage. At least in the short term, some scientists could not keep pace with carefree amateur radio technicians and pushy financiers.

In Strasbourg, Ferdinand Braun had not escaped the pull of business interests. His fascination with the challenge of providing a scientific basis for a new, useful, and reliable branch of technology had led him to produce, just before the end of the century, the high-performance Braun transmitter. In 1898 he had set up the Telebraun Company, and two years later he and the powerful Siemens concern established the Braun-Siemens Company.

In December of 1909 Braun and Marconi journeyed to Stockholm to receive the Nobel Prize for physics, Braun for his transmitter and Marconi for telegraphy. Just a few years later the two would become involved in increasingly heated disputes about dominance, patents, markets, and bases. Marconi worked for the British government, Braun for the German Empire. When war broke out, radio communication, especially in the North Atlantic, became critically important. Both nations had to safeguard their interests in the United States, which was still neutral. The Reich government demanded that Braun, with his prestige as a Nobel Prize winner, represent Germany in U.S. patent courts and elsewhere. Braun traveled from Norway to New York under an assumed name. When the United States entered the war, Braun's assistant was imprisoned immediately. Braun was spared this fate, but he was not permitted to leave the country, and in 1918, he died in Brooklyn.

Ferdinand Braun seems a paradigm of the independent researcher, teacher, and inventor. One would expect him to be fairly well known in his native land. But the five-volume reference work *Great Germans* does not include him. Braun's name also disappeared from the history of the cathode ray tube, now the CRT. His fate as the forgotten forefather of microelectronics mirrors Germany's, in fact all of Europe's, attitude toward that branch of science and its applications. Also forgotten at the time of Braun's death was his hope of using crystals to process electric signals. The enigmatic and unreliable crystal was displaced by the clumsy electron tube. But this was by no means the first crisis in the history of crystals.

Symmetry Solidified

Kryos, the Greek name for frost, is the root of the word "crystal"; ice crystals form when liquid water solidifies. A crystal, a solid body of regular shape, is formed when cooling takes place slowly enough so that each atom has ample time to move to its proper assigned site. Too rapid cooling yields a disorderly mass of crystallites too small to reveal their regular spatial arrangement to the naked eye. Such conglomerates of many very small crystallites are the usual forms of many solid materials, called polycrystalline solids: metals, rocks, ceramics. Only rarely does nature come close to producing a very large "single crystal," visibly displaying the harmonious building pattern of a symmetrical arrangement in space: gemstones, ice, many of the salts. These regular shapes, standing out in the random, disordered world, have fascinated mankind since earliest times.

A liquid such as water yields readily to external forces; it is fluid, yet connected and continuous. It appears to fill space without any discernible order. The crystal, by contrast, is rigid and has an established order in space. Out of the uniformity of liquid, the developing crystal gives preference to particular directions. A regular structure appears.

Snowflakes lucidly reveal the beautiful architecture of crystals. Their needles, columns, and flat plates usually display a six-sided symmetry.

When water in the atmosphere cools to freezing temperatures, crystals form around a nucleus in a dendritic, or treelike, shape. In other solids, such as sand or metal, the building plan of the crystal is hidden from us in the tightly packed mazes of numerous little crystallites. But the pattern is there, and can be seen under a microscope.

Whether one is contemplating the architecture of a snowflake or that of a skyscraper, the form of a novel or a theory of particle physics, symmetry lies at the foundation of our sense of beauty. A motif that repeats or reflects itself, filling time and space with harmony and regularity, is particularly pleasing to our senses. The asymmetry and atonality of modern art forms derive their power from reference to this standard of symmetry and order.

Our quest to understand the phenomena of the natural world is inseparably linked with a search for symmetry and order. From the earliest times, whenever people have discovered or fashioned a rule concerning processes in the physical world, they have felt a sense akin to religious joy, for physical laws suggest a harmony in the cosmos, transcending the apparent chaos. Physicists especially have sought symmetry in the labyrinth of nature; mathematics, that most symmetrical of sciences, has become the window through which modern physicists have glimpsed the awesome universal symmetry of mass and energy, space and time. Our modern knowledge of crystalline solids also requires the mathematical subdiscipline of group theory, the science of symmetries.

To the ancients the natural world was mostly antagonistic, governed by chance and constant change. The courses of rivers, the coastlines of continents, winds, and clouds all appeared unregulated, chaotic. All the more valuable to them, then, were the constant, orderly systems they were able to discover in the material world. One such system was that of the heavenly bodies, particularly the planets, with their constant orbits. Another was crystals—quartz, rubies, sapphires, diamonds, and the other gemstones, whose perfection and incredible symmetry made them beautiful. A system was devised for correlating these crystals with particular heavenly bodies, and both sets of objects were imbued with religious significance.

For the alchemist and physician Paracelsus and some of his contemporaries in the sixteenth century, illness primarily expressed a disruption of harmony. Man, the microcosm, had to be kept in balance with the macrocosm of the heavenly bodies. Medicines derived from crystals, representing the heavenly bodies, could restore harmony between the individual and his world. Powdered crystals, it was believed, could impart strength to the weak and infirm; pharmacopoeias from the Middle Ages have left us a long list of the healing properties of powdered stones. Crystal gemstones became valuable also in the practice of the occult. Priests and oracles used crystal balls to predict the future, and hypnotists induced trances by exploiting the crystal's ability to break up light into a sparkling array of colors.

This mystical spirit of the Middle Ages has made a surprising comeback in the modern America of microelectronics crystals. Quartz and topaz talismans have become important objects of the so-called New Age. Crystal rites are believed to heal the mind and body and to bring financial and personal luck. New Age adherents seem to believe in a kind of Ptolemaic counterrevolution, rolling back modern science to gain unity with the universe on some new evolutionary spiritual path; surprisingly many resort to this new version of a crystal craze.

A Cold Gleam in the Dark

Around 1630 the shoemaker Vincenzo Casciarolo of Bologna must have been inspired by this black magic as well as by the promise of profits to those who mastered it. Bologna's university, Europe's oldest and proudest, was just about to celebrate its five hundredth birthday. Commerce was flourishing, and the citizens of Bologna could devote themselves to the arts and the increasingly exact sciences. Casciarolo, though a shoemaker, devoted every spare nocturnal minute to experiments in alchemy. In his constant search for minerals that promised the unusual, he one day discovered some particularly heavy stones on Monte Paderno that shone in a strange way. He carried them back to his cobbler's shop and began experimenting with them. First he carefully covered the windows so that the neighbors would not become suspicious. Then he heated the stones red hot and added various substances to them, including coal and lime. The new day had almost dawned as Casciarolo extinguished the fire, put his instruments away, and blew out the little lamp. Groping his way out of the darkened chamber, he suddenly saw a pale gleam of light. At first the shoemaker did not believe his eyes, but it was true: his Monte Paderno stones had been made to glow in the dark!

Many more experiments showed that if properly treated with charcoal, those heavy stones would become a cold source of light. Casciarolo recognized that he had found something totally new. Neither the Bible nor Aristotle had mentioned anything about cold light. Of course, there were glowworms and fireflies, but they were alive. The alchemists adhered precisely to the ancient Greeks' differentiation between artificial and natural materials. According to those rules, natural substances could never be created out of something artificial. Artificial light without heat—that was as unheard of as artificial intelligence or artificial life. The clever shoemaker saw a great opportunity in his new discovery: surely he was close to creating gold, the noblest of metals, which belonged to the sun, that glowing star. Because his new substance shone like the sun, he named it his "solar egg." He hoped that from it, great wealth and fame might grow.

What Casciarolo had found was the first evidence of fluorescence and phosphorescence in minerals. He had reduced barium sulfate to a sulfide, but that would not be understood for centuries. In the meantime, researchers racked their brains over the cold light of crystals. Not until the twentieth century did quantum theory unlock the secret of Casciarolo's solar egg. Today we annually manufacture 100 million square feet of luminous screen for television alone. In addition, several billion square feet of glass tubes are treated with luminous crystallite powder to produce more than a billion fluorescent lamps the world over. And it all began with a few stones in a little workshop in Bologna.

Vincenzo told the city's scholars about his discovery. Thanks to Mont'Albano, Licetus, Marsiglius, and others we have the first written documentation—in Latin—of the phenomenon of luminescence. But the scholars approached this crisis for world philosophy very cautiously, beginning their works with long, humble preambles to avoid raising the hackles of the religious authorities. In a more poetic vein the alchemist Scipio Bagatello and the astronomer-mathematician Giovanni Antonio Magini reported on the beautiful new material that drank in sunlight by day and radiated it out by night.

A Collision of Worlds

Northern Italy became the cradle of modern physics. Galileo's experiments started the science of mechanics, which paved the way for a mathematical, quantitative treatment of natural phenomena. Casciarolo and his contemporaries made the first attempts in the branch of science we now call solid-state physics. However, the crystals were so complicated compared to their understanding of matter that those early efforts could not succeed in establishing a scientific approach. The crisis of the crystal was insurmountable. Centuries of development in classical mechanics and then quantum mechanics were necessary to understand solid materials; after Casciarolo's time the study of crystals sank back into alchemy and black magic.

Toward the end of the eighteenth century the natural sciences became increasingly concerned with exact measurements and quantification. Crystals were included in this factual approach. They were carefully measured and described and categorized according to their common characteristics. This cold, categorical approach to substances of mystical importance revolted many thinkers, especially the poet Johann Wolfgang von Goethe. Crystallographers classified their specimens in a totally unnatural manner, Goethe felt. In their search for common denominators, they had come up with the term "isomorphism," meaning having the same shape, and minerals with seemingly different traits were suddenly lumped together. Goethe considered the categories forced; the essence of a blood-red ruby was clearly different from that of an ice-blue sapphire, as anyone could see. The ancients and the alchemists had ascribed radically different symbolic values, effects, and classifications to the two crystals. But the crystallographers, measuring the angles between surfaces and making other calculations, were persuaded that these two disparate stones had the same structure and therefore belonged to the same category.

Today, we know that both rubies and sapphires consist of the chemical compound aluminum oxide. The difference between them lies in the additional elements: ruby has traces of chromium, while sapphire has a touch of iron. The colors of these gemstones are not determined by the innate structures of the crystals, which are identical, but by small amounts of foreign elements. This fact would have been difficult to hypothesize, much less to ascertain, in Goethe's day, when the methods of chemical testing were still quite primitive and could not have measured such minute quantities.

The polemic debates over the merits of crystallography marked the beginning of a great schism in Europe, especially Germany. On one side were the humanists, on the other the scientists. The humanists, dedicated to classical principles, upheld the idea that what matters is the human experience of harmony and wholeness in nature. The lifeless measurements obtained by scientific instruments, the analyzing and categorizing and dissecting of nature, could lead only to dehumanization. In Faust's laboratory Wagner, the unscrupulous, cold physicist, states: "What Nature held as secret, we cleverly analyze. What she organically nurtures, we crystallize." Without batting an eye, Mephistopheles replies that he expects this kind of crystallization to wreak havoc on humanity: "I've already seen crystallized Mankind." Mesmerized, both watch the homunculus take shape in the glowing, steaming vial. Goethe's crystallization is the cold, merciless subjugation to a soulless compulsion. That, he thought, would be the fate of humanity if it relied on science and disregarded intuition, compassion, common sense.

Those who made the effort to understand crystallography could see that a new kind of harmony and beauty was developing on foundations much deeper and more stable than ever before. Unfortunately, the anxieties that arose during Goethe's day have persisted into the twentieth century, preventing Europeans from fully appreciating this new order. It's no great exaggeration to say that the Old World continues to fear that crystals in the hands of technologists will turn human beings into cold, calculating automatons—into the rigid, crystallized homunculus of Faust's laboratory. These attitudes bear a large part of the responsibility for Europe's woeful lag in developing a microelectronics industry.

2

Inner Space

. . . .

SIMPLE geometric forms seem to characterize crystal structure: a magnifying glass shows one that ordinary table salt consists of uniform little cubes. We might say that this structure reveals the systematic harmony that Plato claimed was fundamental to nature. But why does salt have this shape and not that of a sphere or a pyramid? In Goethe's day this question was impossible to answer. Also puzzling was the origin of the beautiful, large, and symmetrical quartz rock crystals, which appeared to have somehow grown like an organic substance. Such structures seemed to be a link between the dead world of minerals and the living world. As scientists continued to study the surroundings of these beautiful symmetrical minerals, they learned that they had been formed by an extraordinarily slow process of solidification. When the liquid material solidified more quickly, the result was a jumble of independent crystallites rather than a single, large symmetrical form. If the formation process was slightly disturbed, "twins" appeared, mirror-image quartz crystal brothers standing side by side.

As methods of chemical analysis improved, crystallographers hoped they would soon find the key to the structure of these solids, but their hopes were premature. Some chemical compounds, such as the oxide of titanium, did not even seem to be able to decide on one particular crystal structure; they might have two, three, or even more structures. Substances that were closely related chemically often had similar crystalline forms, but there were no universal guidelines. And some materials that were quite different chemically evinced the same crystal structure. For a time crystallographers could only doggedly collect and count the minerals; they could describe but not explain them. No wonder Goethe had seen neither rhyme nor reason in their work. What could you expect, from such an artificial approach to the wonders of nature?

Crystals Capture the Light

Observing that light beams travel in straight paths and cast sharply delineated shadows, Isaac Newton suggested that light consisted of particles. Many experimental findings of his day fit this hypothesis, but as the experiments became more refined, they raised doubts. Light passing through a series of narrow slits produced bright and dark patterns so similar to the interference patterns of water waves that the suspicion grew stronger that light was in reality a mysterious form of wave motion. The peak of one light wave, when superimposed on the valley of another wave, results in mutual extinction, meaning darkness. When two peaks come together, the bright portions of the interference pattern are produced. No theory of light as particles could explain this fact. Thus began a long and fascinating search for an understanding of the nature of light.

Crystals soon played a decisive role in the investigation of light and its component colors. Particle proponents explained the color variations of light as a result of the different velocities of the light particles; wave theorists believed that colors resulted from different wavelengths. When light passes through a crystal prism, it diffracts into the colors of the rainbow, a phenomenon that seemed to promise a decision between the competing theories. But the decision was not an easy one. Indeed, the interactions of crystals and light raised even more questions. Looking back at the long and arduous path of this research, one can even accuse the crystal of having impeded and complicated, misguided and confused progress in understanding light.

Two scientific questions were particularly disquieting. First, why did some crystal specimens so strongly affect the color of light in the luminous afterglow? Even more perturbing, how could light be stored inside a crystal if light was not a particle but a wave? These questions were correctly considered as fundamental by the physicists of the time, but some philosophers warned that the entire approach of the exact sciences might be incorrect.

Goethe was initially appalled that anyone could even want to break natural white light into its separate colors, destroying the harmonious whole. But he himself experimented with color and crystal and examined his rich collection of luminescent crystals from Bologna. In 1810, in Section 678 of his book *Farbenlehre* (The Teachings of Color), he described an important contemporary discovery. He concisely summarized some experiments that had probably been done by his physicist friend Thomas Johann Seebeck: blue light could excite the afterglow of the luminescent Bologna stones, but red light could not. This discovery seems to be the first documentation of the quantum nature of light, but because it was just one in a great heap of other observations, many accidental and obscure, no one at the time recognized its significance. A hundred years later Albert Einstein would receive the Nobel Prize for explaining, using quantum theory, this vital distinction between blue and red light: because the quantum energy of red light is smaller than that of blue light, it cannot provide enough energy to excite a crystal's quantum states. To get to that understanding, much, much more work had to be done between Goethe's day and Einstein's—and crystals impeded progress more than accelerated it.

The worrisome question of how light could be stored inside a solid crystal was more fundamental to the issue of particle versus wave. Particle theorists might have been able to resolve that problem by stating that light particles simply could get stuck inside a crystal and that when they became untangled and left the crystal, it would glow. But particle theory had many other weaknesses, and wave theory soon became the accepted explanation. However, the fact that light was stored inside a crystal was a totally unexplained phenomenon.

Light, traveling at a speed of 186 thousand miles per second, oscillates a billion times during a millionth of a second. Yet somehow this wave, oscillating incredibly rapidly and traveling exceedingly fast, could be stored for minutes or hours inside a crystal, as Casciarolo had first noticed. Experiments had shown that light could go through solid matter as a wave, and the strength lost by the wave in passing through a slice of crystal could be measured. But there was no sensible explanation of how light could be stored for times that seemed almost infinite when compared with the short time of one light oscillation. The crystal presented a major crisis for the understanding of light.

One of the most active physicists of the second half of the nineteenth century, Alexandre Edmond Becquerel, a professor in Paris and president of the Academy of Sciences, was preoccupied by the problem of light and crystals. Edmond belonged to a family of scientists: his father was also eminent, and his son Antoine Henri was to receive a Nobel Prize in 1903. All three Becquerels had access to a large collection of minerals in the French Museum of Natural History, where they could perform extensive studies to solve the dilemma of light and crystal. Edmond devised an experiment in which he cut a small hole in the laboratory window shade; the hole could be closed with a small flap. When the flap was raised, sunlight streamed in and passed through a prism, breaking the rays into all the colors of the rainbow. Each color was separately directed to hit a crystal specimen for a short time. To do this experiment Becquerel had to sit for long times in the dark to accustom his eyes and train them for maximum sensitivity. When he opened the flap, he would close his eyes, then look quickly to see what color of light the crystal emitted after the light source was shut out again. The light's color and intensity were recorded. Today we can still marvel at the careful drawings of these results in Becquerel's two-volume *Light: Its Causes and Effects.* Step by step, each crystal was tested. Reams of data and drawings were accumulated to solve the enigma of cause and effect of light in crystals.

But instead of resolving the crisis, Becquerel's careful examinations increased and enlarged the riddles and inconsistencies. Goethe's observation of the very different effects of blue and red light were clearly substantiated, but the wave theory's predictions that a crystal's storage properties were linked to its refractive power were not. When the data were summarized and compared, it was found that crystals of the same type often showed entirely different afterglows. Their color and intensity seemed to be controlled by small admixtures, often too small to be detected by chemical analysis, of other materials. The unscientific recipes that had been used to prepare luminous crystals were not replaced by science; to the contrary, evidence mounted in support of this alchemy! Ground oyster shells, finely powdered pearls, or a pinch of copper seemed to enhance the storage of light, and Becquerel's studies validated these vague, magic instructions. Despair at understanding the phenomenon echoes through the descriptions in his book.

The crisis for wave theory was acute. Yet the physics community, instead of seeing these contradictions as a major challenge, pushed away the crystal results, declaring them unconvincing and erratic so as to not endanger the otherwise beautiful picture of light as a wave. Crystal physics was evicted from the corridors of exact science. Physics is the art and science of the solvable; problems that cannot be solved are rejected as soon as the essential criterion of reproducible results cannot be met.

Crystals seemed to stand in the way of scientific progress. They were so varied and diverse that they were useless for formulating fundamental physical laws. The experiments that laid the foundations for quantum mechanics therefore deliberately dispensed with solids. Indisputable numbers were required to answer the basic question: How does the strength of light emission vary with the temperature of the emitting body? To answer this question, experimenters in the 1890s constructed a hollow sphere and drilled a very small hole into its wall. Through this tiny opening the researcher could observe the light while the temperature of the sphere's walls was systematically changed. When the results were graphed, reliable curves for light output and color distribution as a function of temperature could be plotted. Max Planck eventually explained the results by postulating that light energy comes in quantum portions.

After quantum theory was established, it became clear that light has a dual nature, that particles and waves are two aspects of the same phenomenon. Quantum theory unifies these properties. Now we know that red light has a different effect on crystals than blue light does because the quanta of red light carry less energy than those of blue light. This difference is today an almost obvious consequence of the quantum picture of nature. The mystery of light storage has also disappeared, because we know that light can be stored as energy inside a crystal and then emitted as light. We now know that the energy is stored in the crystal's individual atoms, especially in those of impurities rather than of the host crystal. The diversity and inconsistency of results from the last century has helped provide direction as modern technologists look for useful properties inside the crystals. The crisis has become the blessing of solid state physics. We now can appreciate how difficult this crisis was and how complicated a detour had to be taken to resolve it.

Roentgen's Rays Disclose Symmetry

Wilhelm Conrad Roentgen's discovery in 1895 of X rays turned the scientific world upside down. In 1901 he received the first Nobel Prize for physics. The world over, physics labs as well as doctors and technologists jumped on the bandwagon to reproduce his experiments. The hidden had become visible! X rays are produced by bombarding a metallic target with fast electrons in a vacuum. Roentgen's initial publications on X rays were so precise and comprehensive that his successors had almost nothing left to write about. Nonetheless, the nature of these rays became one of the hottest topics in physics. Were Roentgen's rays like light? If they traveled in waves, they should have the same characteristics as visible light. It should be possible to make the waves interfere with each other; superimposing the positive and negative oscillations should produce a pattern of light and darkness. But attempts to find such a pattern failed, which meant either that the X rays were really not waves but some sort of particle, or that their wavelengths were simply too short to be detected.

The opponents of the wave theory, in particular William Henry Bragg in England, countered that the new rays behaved like a collection of very rapid particles. They theorized that X rays acted like billiard balls bouncing off the bumpers. Roentgen's rays had thus rekindled the old dispute between England and the Continent on the nature of light.

At the turn of the century, many of the best students were drawn to physics and the natural sciences, and the young German physicist Max von Laue was one of these. He went to Max Planck, who proposed that Laue write his doctoral dissertation on the overlap of waves. The oral doctorate examination covered philosophy and chemistry, and luckily Laue was well-versed in chemistry, which made up for embarrassing gaps in his understanding of crystallography. Knowing that the young man would probably never work with crystals, his examiners condoned his ignorance.

At that time crystallography was merely a descriptive branch of mineralogy, not a science in itself. Students learned a bit about it when studying optics, although conflicting information was simply avoided. For Laue, Roentgen's discovery of X rays was much more exciting than boring descriptions of minerals, so he ignored crystallography. He went to teach at Munich, where Roentgen and the influential theoretician Arnold Sommerfeld both taught.

In 1910 Peter Paul Ewald, seeking a topic for his dissertation, approached Sommerfeld, who gave the candidate a wealth of suggestions. Ewald selected the topic about which Sommerfeld had the greatest reservations, but the professor finally assented. Ewald wanted to investigate whether something oscillating inside a crystal could explain the properties of light passing through it. Becquerel and many others had done experimental studies along these lines.

Soon afterward a historic conversation took place. Ewald wanted advice from Laue, who was now a junior professor. As the two men walked through the English Garden, Munich's famous park, Ewald explained his approach to the problem. Laue listened carefully but at first did not seem to understand what Ewald was getting at. "Why should there be something oscillating inside a crystal?" Laue asked. Ewald explained that many crystallographers had proposed this hypothesis, which seemed to offer hope for understanding and calculating how light travels through a crystal, how it becomes reflected, and how it is absorbed. Something oscillating should act like an antenna. Laue wanted to know what these oscillators could be and how far apart they were spaced. Nobody knew, Ewald admitted.

Laue pressed on: how closely might these oscillators be packed? What separated them? Ewald, irritated, gave a vague answer. "It seems to depend . . ." His vagueness angered Laue, and the discussion became tense. Ewald swept away the question and said that to him these distances did not matter at all; he wanted to construct a general theory. The distance between the oscillators might be one thousandth or one ten thousandth of the wavelength of visible light. Ewald continued on the topic he had come to discuss; he wanted Laue's comments on the prospects for success of the various mathematical techniques he thought useful. But to Ewald's chagrin, Laue did not seem to be listening. Twice Laue interrupted and asked the same question, completely out of context. "What happens to very short waves inside a crystal?" Ewald had to struggle to remain polite: "I do not know about very short waves. I hope to find some general answer, which might include an answer for this special case." As their walk ended, Ewald curtly stated that he had much work to do, closing the discussion with great disappointment.

But Laue could not forget this discussion, for it had given him a totally new idea. It might be possible to link two apparently unrelated problems of physics in a single experiment. If X rays were waves, then they must have very short wavelengths, and if crystals were made up of a regular array of oscillating atoms, then X rays penetrating a crystal should be widely scattered. The scattered light pattern would prove that X rays were waves and that crystals consisted of regularly spaced atoms. Crystals, the black sheep of physics, and X rays, its darling, might have something in common.

Interference patterns are a sure proof of the wave nature of radiation. Such patterns occur whenever two rays are made to travel along paths that differ slightly in length, so that at a detecting screen the two rays are no longer in step but are slightly shifted in relation to each other. If a wave maximum is superimposed on a minimum, all light is canceled. Only when the length of the path is an exact multiple of the wave length of the ray does one see evidence of the radiation in the form of a bright spot of light. In that case the two rays are said to interfere constructively. Such interference patterns can be obtained when some regularly spaced obstacle breaks up or diffracts the waves into separate beams, which then are brought together again. The spacing of the intercepting obstacle must be of the same order of magnitude as the wave length of the radiation. Finely ruled lines on a glass plate, called gratings, can do this job for visible light. But how might one find such a regular spacing for a very much shorter wavelength? The most that could be deduced from experiments thus far was that X-ray waves might be as little as one thousandth the length of waves of visible light. That was why Laue had suddenly fallen silent during his talk with Ewald. He had been caught up in pursuing these ideas!

If Ewald's theories about the architecture of crystals were at least partially correct, then crystals might provide the right match for X rays. The regular spacing of the atoms in a crystal might serve as the needed regularly spaced obstacle to diffract the X-ray waves and thus create interference patterns of dark and light. X-ray wavelengths and atomic spacings inside the crystal might luckily coincide to reveal an interference pattern. This was Laue's hypothesis.

Walther Friedrich, who had just received his doctorate under Roentgen's tutelage, was now Sommerfeld's assistant and wanted to pursue Laue's ideas. But only after Paul Knipping, another of Roentgen's doctoral candidates, was recruited, did the experiments begin. The researchers carefully set up a blue copper sulfate crystal, near a balloon-shaped X-ray lamp shielded from the crystal by a thick lead plate. Through a tiny hole in the plate they focused the beam of X rays on the crystal. They correctly assumed that only a small portion of the X rays would be diffracted, that most of them would travel unimpeded through the crystal. A piece of lead shielded this strong undiffracted beam from reaching the photographic plate, which was meant as a detector for the interference pattern Laue had predicted.

The first experiment was a failure. The second, however, clearly showed individual dots of light in a fairly regular arrangement. They had found the pattern of light and dark! But Roentgen, a semi-recluse by this time, was not impressed by the experiment, and Sommerfeld had already voiced his objections. Laue had resorted to such terms as "optical feeling" and "instinct" to hypothesize that some kind of refraction had to occur, but he had not been able to predict what the pattern would look like. On his way home after the second experiment, Laue pondered the photograph that showed the wreathlike pattern of light and dark. He had almost reached his apartment when he suddenly, instinctively, knew what the mathematical formula had to be! He already knew how light is diffracted at an optical grid, which is a tightly packed row of fine lines. If the crystal were a systematic, three-dimensional lattice, he would have to write and solve the equations three times for the three dimensions. Next he would have to interpret the discovery. His wreath of glowing dots had to rest on cones in space; moreover, the pattern had to be determined by the three conditions of diffraction. The theory was still incomplete, but it covered the essential points and was mathematically solvable. Using the unambiguous language of mathematics, Laue could present his findings to the scientific community. Researchers had shot X rays through minerals before, but no one had thought to look for the tiny fraction of the X rays that did *not* go straight through but was deflected. That experiment had required Laue's theoretical prediction.

Laue decided to reveal his discovery in Berlin on June 8, 1912, in the same auditorium at the University of Berlin where Max Planck had first reported on his quantum theory twelve years before. In his letter announcing his arrival, he enclosed a photograph of the dot pattern and offered a prize to anyone who could guess what the pretty arrangement meant. Of course, no one was able to claim the prize.

Laue's hypothesis turned out to be correct. Many others reproduced his experiments, including William Henry Bragg and his son William Lawrence, who supplemented and simplified the results. Crystals had finally mastered a full-scale crisis, permitting solid-state research to come into being. Scientists now knew not only that X rays had wave characteristics, but also that their wavelength was about one billionth of a centimeter.

More important, Laue's finding allowed scientists to study the crystal's interior. Precise numerical interpretation of the diffracted light beams enabled researchers to make concrete statements about the location of atoms in the crystal. X rays established the positions of the nuclei and revealed how the electrons surround them. One could now measure in minute detail the electronic bonds between the atoms as well as every disturbance and irregularity in the systematic structure of a crystal.

Regularly spaced atoms are the building blocks of crystals, and their electron clouds are the oscillating entities that Ewald had assumed. The spherical distribution of electrons around sodium and chloride atoms explains why table salt has a simple cubic structure. The more complex distribution of electrons in other materials, such as quartz, results in a more complicated crystal structure. Chemical bonding and crystallographic structure were now understood as being intimately linked. The rigid tetrahedral alignment of electrons in carbon atoms forms diamonds. Chemistry, the study of atomic bonds and compounds, was no longer limited to the study of liquids and gases and was finally linked firmly with physics and crystallography. With crystals, scientists at last had access to an understanding of solid substances.

X-ray interference patterns had disclosed the symmetry of crystals, once again showing the close relation between the harmonious beauty of crystals and the mathematical concept of symmetry. With X-ray measurements it was possible to determine for each crystal structure all its elementary forms of symmetry: rotation symmetry around an axis, mirror-image symmetry with respect to a mirror plane, and translation symmetry. These properties of the regular spatial arrangements of the atoms determine some of the elementary physical properties that a crystalline material can have. The awesome complexity of describing such a large number of atoms by a mathematical theory was now substantially reduced by the concept of symmetry. One could now have hope for a true quantum theory of solid matter.

Research on crystals underwent a major change; the trial and error approach, characterized by amateurish tinkering and unguided observations, was superseded by the scientific method. Predictions could be made and tested with mathematical accuracy, and the results could be analyzed in terms of hard facts and precise numbers. The era of amateurs—poets or naturalists—was gone forever. Crystals were now admitted to the rigid world of physics. When Laue first unveiled the crystal's secret, only a few researchers showed any interest. But in 1914, at the start of World War I, Laue was awarded the Nobel Prize.

Individual contact with a great teacher remains one of the most important spurs to a scientist's personal performance and to the progress of science. In his memoirs Laue mentioned the impact of Ferdinand

Braun's stimulating physics course at Strasbourg University. Laue had to obtain leave from his service in the Prussian army to attend the lectures; Braun, in his memoirs, described how the eager student usually arrived late, in uniform, helmet in hand, disturbing the class as he found a seat. Braun's teaching laid the foundation for Laue's later work. And I, in turn, am thankful for having met Max von Laue, who influenced my methods and directions. After World War II, he was appointed to revive and head the Kaiser Wilhelm Institute for Physical Chemistry in Berlin, and in 1950 he paid his first visit. At that time I had received my high school diploma but could not yet enroll at a university. After completing training as a locksmith, I managed to get a job as a lab assistant at the Kaiser Wilhelm Institute. Our department, which was studying light emissions from crystals, was feverishly awaiting Laue. He was already long overdue, having gotten caught up in conversation somewhere else at the institute. We waited and waited until one o'clock. The timing was critical, because cafeteria meal tickets were good only until one-thirty. So everyone went to eat except me—the lowly lab assistant who did not have the privilege of meal tickets. I sat on a lab stool and ate my sandwich alone. Suddenly the door opened and in walked Laue, with the administrative director in tow, who desperately tried to persuade Laue to return later. But Laue insisted on having the youngest member explain all of the instruments and projects to him. Unexpectedly honored, I enjoyed his friendly, supportive comments.

Years later I was able to thank Laue on behalf of the German Physical Society; I requested that the German Postal Service issue a stamp commemorating his hundredth birthday. The stamp portrays a "Laue Diagram" in all its glory: illuminated X-ray dots disclosing the secret of the interior of a crystal.

Diamond Chips and Table Salt

Crystals were paving the way for microelectronics. In Göttingen just after World War I, the experimental physicist Robert Wichard Pohl, an excellent university teacher, wanted to pursue nuclear physics, the number one topic of the time. He planned to begin by researching how electrons were emitted when light was beamed at atoms near the surface of metals. The investigation required glass vacuum tubes, and Pohl was angry when he realized that in postwar Göttingen he was unable to obtain liquid air, which freezes water vapor and makes it easier to achieve a vacuum. At first, he joked that they would just have to start at the other end. If they could not use vacuum tubes, they would have to move the atoms under study inside the metallic crystal, where they could observe the photoeffect. The punch line was that this joke, made in desperation, established a new direction for solid-state physics!

To study photoconductivity, that is, electrical conduction stimulated by light, Pohl and his assistants needed just the right sort of crystal. Powdered phosphorescent zinc sulfide sometimes worked well but was not always reliable. They procured a specimen of diamond, the noblest of crystals, and began experimenting. To ascertain the exact number of electrons moving as a current and to obtain precise information on their movement, a scientist makes use of the fact that a magnetic field will deflect the electrons from their paths. This principle of measurement is known as the Hall effect. Pohl and his helpers carefully clamped their diamond in a large electromagnet, secured all the leads, and adjusted their lamps and filters. But as fate would have it, they forgot to screw down the heavy iron poles at each end of the magnet. When everything was ready, Pohl ordered the magnet to be switched on, but instead of the deflection of electrons, all they got was a shattered diamond. The heavy magnetic poles had violently slammed into each other and pulverized their noble crystal!

Pohl was furious at his stupidity, and as a result, stopped using this important technique. After the diamond fiasco, the Göttingen group continued to search for the best family of crystals to use in reproducible, convincing experiments. Diamonds were too expensive, so the researchers resorted to table salt and its closest relatives. They could afford sodium, potassium, rubidium, chlorine, bromine, and iodine, whose crystals had simple cubic structures.

The Making of a Crystal

Pohl soon realized that impure crystals from mineral collections were undependable. Many earlier researchers had been frightened off by the nonreproducibility of experiments using natural mineral specimens. So he established a laboratory to grow his own large crystals from pure raw materials. His procedure became a model for scientific research, and today's labs still use similar methods to grow large crystals for microelectronics.

To grow a crystal, very pure, chemically cleaned powder of a salt compound is poured into a crucible, which is then heated in a furnace. After the material reaches the melting point, a small crystal piece, selected for its high degree of perfection, acts as the "seed." It is slowly and carefully dipped into the melted material. As the tip of the crystal cools, the melt solidifies. If the process occurs slowly enough, every atom moves into its proper place; the mathematical ideal of symmetry becomes a solid reality. The crystal is then carefully removed from the melt; the temperature must be controlled and the process of solidification gently monitored. This is how humans create a large crystal; nature takes eons to do the same, but cannot approximate this degree of purity.

The artificial crystal put an end to years of uncertainty. The old dispute between natural and artificial materials had developed a new facet: scientists could now routinely produce perfect artificial inorganic substances. The "monocrystal," or single crystal, had two big advantages over the natural type, which usually consists of numerous tiny crystallites jumbled together. First, because a monocrystal is regular and symmetrical over its entire length, an atom at one end is a precisely prescribed distance from an atom at the other end, a mathematical ideal. The second advantage of artificial single crystals is their purity. Depending on where natural crystals are found, they may contain, for example, more iron or more aluminum. Such contaminants distort the symmetrical lattice. One area may contain only a few foreign atoms, while many impurities may be clustered together nearby. It was the impurities, as it turned out, that caused all the reliability problems with crystal receivers, with which one found a suitable spot just by chance. Artificial crystals alleviated these difficulties.

Today's microelectronics industries routinely produce silicon crystals with only one or two impurities—at most—per billions of silicon atoms! Every test conducted with pure crystals has opened new doors. It has been found that many characteristics, such as color and electrical conductivity, are not inherent properties of the crystals, but almost always caused by foreign elements. Those tiny admixtures are the determining factors. Scientists have learned to create very different properties in a crystal by adding minute amounts of a foreign substance; "doping" is the term for this important technique. In the past they had to struggle with a variety of unpredictable phenomena, whereas now they expect to control and adjust them. One need only produce a pure crystal and then dope it correctly.

It was also found that a defect in lattice structure could change a crystal's properties. When Pohl's team removed a negative chlorine atom from table salt by carefully heating the salt in a vessel containing metal vapor, a negative electron replaced the chlorine ion, whereupon the previously colorless salt crystal turned deep blue. Studies of color in crystals became an important area of research. Scientists finally proved that any deviation, any defect in a crystal, could completely alter its characteristics. They at last understood why crystals are so sensitive to external influences. Bit by bit the electronic structure of these "color center" defects became unraveled. When each atom was at its expected site in the spatial lattice, no surprises would take place. If, however, a site remained vacant or was occupied by a foreign atom, then color, magnetism, electric conduction, or photosensitivity might be observed. The storage of light, inexplicable in the preceding century, was also found to be caused by defects. It was a lack of appropriate equipment that spurred Pohl on to unlock some of these secrets of solid matter! He accomplished his original plan; the observation of atoms was no longer restricted to bulbs and tubes. Scientists could introduce guest atoms into host crystals, assign them their proper places, and study the results precisely and inexpensively. "Solid-state physics" was about to become a respectable branch of science.

Crystals and Quanta

Around 1930, with the now firmly established quantum theory, physicists could also adequately explain the long-standing enigma of how light is stored inside a solid crystal. The mystery of the Bolognese stones and the irritating contradictions of nineteenth-century physics had not yet been resolved. Quantum theory unified the pictures of light as a particle and as a wave. The experimental condition reveals either one of those two aspects. Light enters the crystal as an electromagnetic wave, but a foreign atom in the crystal perceives it as a photon whose energy is determined by the color of the light. An electron of the foreign atom can utilize this photon's energy to jump to a state of higher energy. Under favorable conditions the electron can remain at that state for long periods, storing light. When the electron jumps back to its original state, the light is emitted. Some of the initial energy may be paid as a "tribute" to the host, in which case the light emitted may be green, yellow, or red, even if the initial beam was blue. Some of the remaining energy is transformed into oscillations of the crystal's atoms.

At first, it seemed unthinkable that light could act both as a wave *and* as a unit of energy, yet these two characteristics proved to be united. This new branch of physics aroused the interest of scientists the world over, including Nevill Mott in England and Frederik Seitz in the United States.

Theoretician Wolfgang Pauli's research on the quantum theory of atoms was so important that a theory was named for him. The Pauli Principle states that at any one time no two electrons in an atom can exist in states defined by the same set of quantum numbers. This axiom helped clarify the structure and chemical properties of the atom. Pauli elucidated the elementary magnetic characteristics of the individual electron, which also helped explain magnetic relationships in metals. Pauli refused to let his students work with solids, but Rudolf Peierls tried to do so anyway. When he sent his teacher a manuscript on the electric conductivity of metals, it was returned to him covered with scathing remarks: "One shouldn't dig around in garbage!" "Cast-iron physics is out of the question!" The young man could certainly find better things to work on! The student should be thankful that the Master had not torn the manuscript to shreds!

Quantum theory applied to atoms and their constituents, the elementary particles, remained at the center of physicists' attention. The theory was able to provide deep insights into the structure of matter and serious philosophical conclusions concerning our perception of the universe. In contrast to this pure scientific approach, learning about the physics of solid materials seemed a lowly technical task, still full of chance and circumstance, dependent on accidental impurities and involving far too many particles to allow any mathematical predictions of physical properties. Solid-state physics began as a Cinderella. It was looked down upon as impure, but its time of glory was to come.

The famous theoretician Werner Heisenberg, for one, was not convinced that the solid state could truly be a subject for basic research. I had attended some of Heisenberg's lectures in Göttingen, and shortly before his death, I talked with him in Munich. We discussed the preliminary planning for a new institute for solid state research of the Max Planck Society. He was melancholy and skeptical about establishing such an institute: was it the right thing to do? Wouldn't people there simply reiterate classical theories? How was anything new supposed to result?

Such skepticism from an outstanding researcher must have had a dampening effect on the course of German solid-state physics. After World War II the young Federal Republic of Germany showed much less interest in crystal research than did the United States. And in Italy, where Enrico Fermi was the nuclear physicists' idol, the emphasis lay in other areas. Whenever I now read reports on the Stuttgart researchers, I remember Heisenberg's doubts and resignation. Yet I am sure that recent results in solid-state physics would have convinced him of their importance, especially Klaus von Klitzing's Nobel Prize in 1985 for work on quantized electron motion in a silicon crystal, work done in the same institute which Heisenberg once regarded so skeptically.

Even in the 1950s in Germany, theoretical and experimental physicists continued their crass, mutual distrust of each other. Candidates for a doctoral degree in experimental physics nearly had to resort to cloak-and-dagger schemes to attend a theoretical physics seminar, a contrived schism that was also responsible for Germany's later lag in microelectronics. In the English-speaking countries the picture was totally different; researchers promoted the unity of physics, and the study of solids.

Another, more important reason for the separation between applied and theoretical research was that applied research could be directed toward political ends. Pohl's team had chosen salts as the subject of research not only because of their convenience, but because they had no applications for industry or the war effort. External factors would not disrupt the research, and valuable time need not be spent applying for patents, which delays scientific reporting and burdens the open exchange of results.

Pohl had good reason to mistrust use-oriented research. During World War I, he had had to work on radio waves and transmitters, and in World War II, war-oriented studies contradicted the essential spirit of free scientific research. Military hierarchy rarely goes well with science. Physics has never demanded that junior members or those lower in the hierarchy obey blindly. On the contrary: young physicists often can provide a new impetus.

Göttingen was the hub of new physics in the 1920s, although it remained a provincial town. No industrial research lab was nearby, and Berlin, the capital of electronics, was far away. The all-important exchange of ideas between industry and academia did not take place. Only many years later, in America, did a new form of teamwork arise: cooperation between industrial parks and nearby universities.

It would be unfair to chastise the German professors for their attitudes, for they had fought courageously to keep scientific research independent of the state, religion, and big business. But the wave of industrialization at the end of the nineteenth century and the rise of the new technologists endangered the prestige of the professors. Technology, they feared, would jeopardize cultural values. Those in the humanities waged war on both technology and the natural sciences, and the battle raged most violently in Germany.

The physics faculty was still part of the liberal arts tradition at German universities. A full professor enjoyed a high degree of prestige, for he exemplified pure scientific research. He was very well paid and his job was extremely desirable. To seek an alliance in the industrial world would have been a disgrace. One did not endanger one's career chances by mixing with industrialists. Universities did not train physicists for industry, nor did they dare run their own businesses.

In chemistry, in contrast, it was easier to move from university laboratory to factory; the relationship between the two has always been much closer. Chemists and chemical engineers have thus always occupied the highest positions at large industrial firms. And Germany, which has won a great many Nobel Prizes for chemistry, still maintains a dominant position in the international chemical field.

As the first third of the twentieth century came to an end, the study of crystals had brought about two major victories. Scientists had convincingly demonstrated the crystal's systematic spatial arrangement and measured the details of its structure. Second, researchers now realized the wide-ranging impact of every deviation from the ideal symmetrical norm. The centuries of often unsuccessful research on the enormously confusing variety of phenomena in solids had not been in vain. On the contrary: a whole new set of challenges awaited mastery by imaginative advances in technology. The crystal had apparently shed its mystical veil; the magical link with cosmic harmony had been demoted to a predictable three-dimensional structure. The stage was finally set for modern solid-state physics. The international brotherhood of researchers was ready to compete peacefully in this new field.

This competition did not take place without suffering and tears, however. In the 1930s Jewish scientists were denounced and cast out of a Germany they had loved, their valuable contributions to science denigrated. Those who stayed behind were crippled by shame and disgust, and their well of inspiration dried up. As the tragedy marched forward relentlessly, the crystal had another crisis to face: compulsory service in the machinery of war.

3

Under the Radar Umbrella

AFTER British forces retreated from Dunkirk in 1940, Britain faced bankruptcy and the loss of its ally, France, recently occupied by the Germans. In May 1940, it looked as if Germany and Italy would conquer Europe while Japan took the Far East. The United States steadfastly refused to intervene. Not until March 1941 did the U.S. Congress enact the Lend-Lease Act, providing support for Hitler's enemies. During the next four years, more than twenty-five billion dollars flowed from America to Great Britain. British scientists could now conduct research that would aid the war effort. Of particular importance was the refinement of an electronic system for detecting the position of moving objects at great distances. Radar (a contraction of *radio detection* and *ranging*) was destined to change the course of the war.

Scanning the Skies

In the 1930s the Scottish physicist Robert Watson Watt had discovered that when short radio waves are transmitted, metal objects will echo them back. This interchange requires sufficiently strong transmission and sufficiently sensitive receivers to pick up the weak echoes. An important aspect of the discovery was that radar could be used to determine the distance between transmitter and object. Since radio waves travel at the speed of light, one has only to note the time lapse between transmission and recovery in order to ascertain the distance.

Initially, Watt intended this new technique for weather observations and forecasts, but the British military quickly realized its strategic importance. A continuous chain of radar stations around the British Isles would be able to detect any approaching aircraft. A "Royal Radar Establishment" was quickly set up to develop the techniques, and large vacuum tubes were constructed to generate the radio waves. By 1939, when the war broke out, enough stations had been set up to form a radar curtain that would alert the country to the approach of any airborne invader.

In July 1940 the Germans began their massive air raids on England. Supported by their air force, the German army planned Operation Sea-lion to cross the English Channel and invade the country. The Germans' numerical superiority made this victory seem likely: Germany had 2,500 bombers and fighter planes, the Royal Air Force had 900 fighter planes. Germany had all the strategic advantages as well: plenty of room on the Continent for airfields and a submarine fleet already surrounding Great Britain. Despite the odds, the British clearly won the Battle of Britain of October 1940. The Germans couldn't compete with Britain's full-scale use of radar electronics. The British victory heralded a new, scientific defense.

The United States and Great Britain poured an increasing amount of money into radar technology, pushing it to the forefront of war-related scientific research along with the Manhattan Project, which developed the atom bomb. Every fifth physicist in America was summoned to work on radar. Gigantic laboratories sprang up in the traditional centers of learning on the East Coast: Massachusetts Institute of Technology established a large radar lab; Harvard and Columbia feverishly worked on short radio waves. The researchers soon established close contacts with the electronics firms that were recruited to produce the new instruments; 2,000 radar units were manufactured each month. Huge antenna sets were built to make precise range determination possible. One type of set, with large round antennas, looked like Mickey Mouse and was soon affectionately called by that name. As reception techniques became increasingly refined and sensitive, scientists learned to use ever shorter waves. The shorter the wave, the farther away and smaller the detected object can be.

Teachers, technologists, businessmen, and politicians worked together in harmony, and the radar program grew by leaps and bounds. There were a number of reasons for this. Labor was abundant, and the American economy, on the rebound from the Great Depression, was ready to spawn a new industry. The number of Americans employed in electronics rose from about 110,000 before the war to about 560,000 in 1941. The electronics industry as a whole grew almost twelvefold.

Adding impetus to this activity was the patriotic zeal of research scientists. Many of them had heard shocking reports of Nazi tyranny from Jewish immigrants, and still more felt the threat to the United States posed by the Axis powers. Scientific organizations and universities drew up resolutions promising President Roosevelt their undivided support and pledging to postpone work on projects unrelated to the war effort. In September 1940 one of the leading engineers at General Electric, L. A. Hawkins, stated, "The Nazi juggernaut was created by engineering and it is only by engineering that it can be destroyed . . . research and engineering must work together."

In June 1941 the federal government created the Office of Scientific Research and Development, which brought together famous physicists and engineers from around the country, particularly from the East Coast universities and Berkeley and Stanford. Government-sponsored researchers worked together closely and also exchanged information with their British colleagues. Scientific research, traditionally the province of universities and individuals, now became the domain of state and federal laboratories. Scientists who worked on those programs still speak of the incredible sense of urgency uniting them against an inhuman regime. Their commitment to the Allied cause was extraordinary at a time when most Americans were reluctant to enter the European conflict.

In Germany, on the other hand, engineers were working on more conventional weaponry like missiles and jets. The German government had not developed a spirit of cooperation with the science establishment to explore the frontiers of technology such as radar. What radar they had was limited to the use of longer radio waves, which were easier to generate and to measure but much less sensitive in detecting objects. The Nazi government mistrusted the meddling of scientists in warfare. When one of the first radar sets on an airplane was demonstrated to Goering, he replied that he did not want "movies on a plane." Goering, who had been a fighter pilot in World War I, felt that the Germanic warrior should face the enemy without the aid of technical instruments. He and many others still believed in Teutonic hero mythology, in which a cool head and a brave heart always triumphed. Third Reich leaders wanted "German physics" that could be readily understood by the average citizen. Ignoring the scientific frontier, military leaders preferred to bet on speedy attacks and blitzkriegs, taking advantage of their numerical superiority in conventional weapons.

Germanium Rediscovered

On a single day in September 1940, 185 German fighter planes detected by British radar were shot down over England. That day decided the outcome of the battle for England. Ironically, a previously unimportant element named germanium helped the British triumph over the Germans.

In 1886 the German chemistry professor Clemens Winkler, researching properties of minerals, arrived at a critical point. When he analyzed crystals of argyrodite, he found that the sum of its component elements, silver and sulfur, did not add up to the correct total. There had to be another element in the crystal, one that had not yet been identified. Or had it? Fourteen years earlier the Russian chemist Dmitry Ivanovich Mendeleyev, in his periodic table of all the chemical elements, had left an empty space below silicon. Mendeleyev had postulated an unknown element for this slot and had derived its approximate properties. Winkler made the connection and showed that the unknown element in the argyrodite crystal fulfilled all the requirements for the unoccupied slot in the table of elements. The discoverer of an element is entitled to name it, and at that time of European nationalism a German patriot could only choose the name germanium, to offset the discovery by French scientist Lecoq de Boisbaudran ten years earlier of the element named gallium.

Research physicists immediately went to work on germanium. One of its properties was that it sometimes acted as a conductor of electricity, but it was not reliable. No two specimens of germanium seemed alike in their conductivity. This new element simply did not seem to be a true representative of the class of metals. The crystal's strong reaction to minute traces of other elements frustrated scientists' attempts to properly categorize it. Despite its quirks, empirical trial-and-error studies indicated that germanium was suitable for use as a crystal receiver for radio and radar waves. Researchers pressed a fine "S"-shaped needle onto the germanium crystal to create a springy, elastic contact, then encapsulated the crystal and the needle tip in a small glass tube with two little metal legs as electrical leads linking the crystal to the outside world. After many exhaustive experiments, this detector was able to pick up even weak radar waves with very short wavelengths. Numerous germanium rectifiers were put to work, although the crystal's mode of operation was still unclear.

Thus the advent of radar in World War II revived the techniques Ferdinand Braun had experimented with in Leipzig. Detecting short waves requires a fast-reacting instrument. The hard part is to create a receiver that can truly track the rapid oscillations and not merely indicate a useless average value. The structures must be small—thus, microelectronics. The incident radio wave should not put too many electrons in motion, and these should be able to rapidly flow through the lowest resistance possible. Braun's point contact fulfilled these exacting requirements; the British used the "cat's whisker" contact, which they praised—but also damned—for its sensitivity.

Semiconductors Take Center Stage

The sudden use of germanium brought to the limelight a suppressed stepchild in the family of materials: the semiconductor. The first two syllables of its name imply something half baked, unfinished, perhaps valueless. The semiconductor bears its name because its characteristics put it somewhere between electrical conductors, such as metals, and nonconductors, or insulators. Metals put up very little resistance to electric current and also conduct heat well. They are opaque, reflect light, and have shiny surfaces. Insulators, in contrast, are usually transparent and do not strongly reflect light. Members of this class, including quartz crystals, glass, and salt crystals, usually conduct both electric current and heat poorly.

This differentiation was not so problematic in classical physics. For reasons not fully understood in the nineteenth century, metal crystals are distinguished by the mobility of their electrons. When atoms join to form a metal crystal, each atom donates one, or at most two, of its electrons, to the total structure. In this theory a metal crystal consists of a systematic arrangement of atomic cores in which most of the atom's electrons are held fast, and a common electron "soup" pervades the entire crystal lattice. These common electrons have no permanent address, so it is easy to see why metals conduct electricity so well. The movement of electrons to the positive pole is called the flow of current. Heat is conducted well because the electrons can absorb thermal energy at a warm area, transport it, and release it at a colder area. Light cannot penetrate this kind of crystal; the many highly mobile electrons react to incoming light by oscillating in resonance with it, thereby emitting radiation, and thus reflecting the light.

Nonconductors act in a totally different manner. When their atoms assemble, they cannot donate electrons to the crystal as a whole. Each atom needs its electrons to form chemical bonds with its nearest neighbors and thus to build up its particular crystal structure. There are no free electrons to oscillate in unison with a light wave and thus reflect the light as metals do. The absence of freely mobile electrons also results in the nonconduction that is typical of insulators, such as mica, quartz, or marble.

The difference between the two categories appeared clear-cut, although for a long time it was not clear why some elements were metals and others were insulators. But more exacting investigations of the electrical conductivity of materials revealed problems. Physicists found more and more crystalline minerals that refused to fit into either classification. This irritating group of misfits was dubbed the semiconductors. Some semiconductors are grayish in color, almost like metals; others are brightly colored or colorless and transparent; no simple criterion seemed applicable.

Semiconducting crystals, such as the elements selenium, germanium, silicon, and boron, and chemical compounds such as lead sulfide, copper oxide, and cadmium selenide, soon displayed their fickle nature. Conductivity apparently varied from experiment to experiment and specimen to specimen, and even from time to time! A proper metal, even if it is not totally pure and clean, evinces a constant conductivity value, which the electrical engineer can gauge precisely. Conductivity values for semiconductors, however, show a wide range. When scientists tried measuring not only the crystal, but also its chemical surroundings, they were surprised to discover that the amount of moisture in the air made a big difference in conductivity. Because of their peculiar sensitivity, semiconductors were soon dropped from proper physics and dependable technology. It was best to steer clear of them.

Some researchers, though, continued to investigate these minerals. Michael Faraday had found that heated crystals conducted electricity better, which was surprising, for heat makes metals *less* conductive. When metal is heated, the atoms in the crystal begin to oscillate around the point where they should actually be sitting. The disordered oscillations of the atoms make the metal expand (as we observe of railroad tracks on a hot summer day). When an electron, the current carrier, moves through hot metal, it finds it increasingly difficult to avoid the vibrating atoms in its path. Heat impedes the electron's progress and lessens the ability of the material to conduct current. Those obstinate semiconductors, though, do just the opposite; their resistance decreases as they are heated. Semiconductors are also extremely sensitive to light, as the Becquerels had realized. Some crystals act as insulators when they are in the dark, but if a light is turned on they behave like metals. They react to small changes of pressure and tension as well as to magnetic fields and surrounding gases. Such irregularities did not conform to the rules of classical physics.

Before quantum theory, the structure of atoms in a crystal was still beyond description. Bit by bit, though, quantum theory revealed the structure of atoms as individual configurations, each with a nucleus surrounded by electrons. Theorists then gradually tackled the greater complexities of solids. Quantum theory poses strict limits concerning the energy values an electron can assume. Each energy level of an atom can be occupied by no more than two electrons, which must differ in spin. This rule continues to be valid when the atoms are tightly packed and their electron clouds begin to overlap. The tight crystalline packing and the electronic overlap change the very sharply defined energies of the allowed electronic states into broad bands of allowed energies separated by energy bands that cannot be occupied by electrons (Fig. 1).

This fundamental conclusion from quantum theory, named the "band model," was a major step forward in determining the electronic structure of atoms in a crystal from the electronic energies of a single atom. In

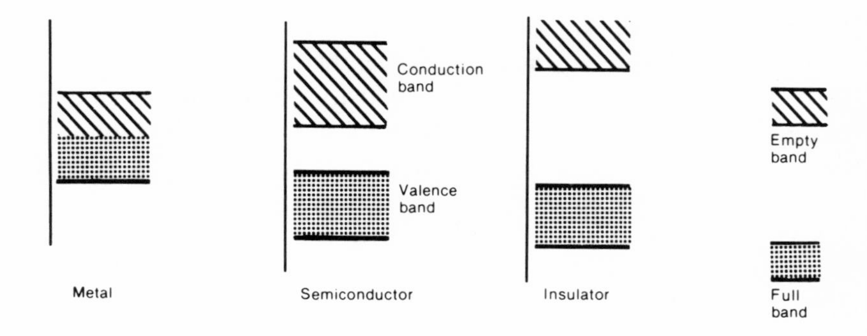

Figure 1. Energy bands in solids.

theory, calculating and predicting electronic properties should have been straightforward, but it was not easy in fact to carry out the calculations. The amount of arithmetic was too formidable prior to the advent of electronic computers. When large computers became available the theories could be tested through comparison with and interpretation of experimental measurements.

In the early days of research on semiconductors, many rather awkward problems remained. For example, when some materials were exposed to a magnetic field, they reacted in unexpected, almost crazy ways. At Johns Hopkins University in 1879, American physicist Edwin Herbert Hall had discovered something new about magnetism and electricity. Hall passed a current from a battery through a piece of gold leaf, then attached contact leads to the edges of the gold leaf and noted, when he switched the magnet on, that a voltage developed (see Fig. 2). "The experiments were hastily and roughly made," he stated, "but are sufficiently accurate . . . to determine the law." The "Hall effect" occurs because the magnetic field deflects a moving charged particle sideways; a charge imbalance is built up that is measurable as voltage across the conductor. This phenomenon became important in the analysis of electric current flowing through solids.

In most solids, such as Hall's piece of gold leaf, the sign of the voltage indicated that negatively charged electrons were carrying the current, as expected. But in some materials, especially those unreliable semiconductors, careful measurements seemed to indicate particles with a positive charge. This "anomalous Hall effect" remained unexplained for many years until quantum theory showed that missing electrons in the semiconductor—"holes" in a sea of electrons—move in the opposite

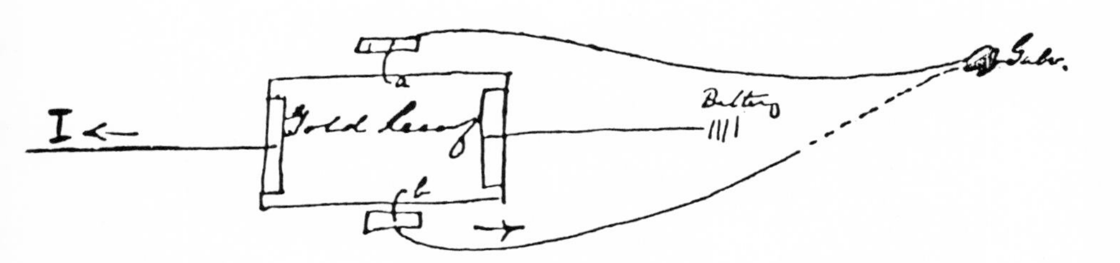

Figure 2. Edwin Herbert Hall's sketch of what became known as the Hall Effect. The current, *I*, supplied by a battery, flows through a gold leaf conductor. When a magnetic field is set up perpendicular to the gold leaf, voltage can be measured between points *a* and *b*.

direction of the stream of electrons, producing a Hall effect as if positive charges were flowing. (A hundred years later the Hall effect once again was news when Klaus von Klitzing received the Nobel Prize for Physics for discovering sharp quantum jumps in the Hall effect of semiconductors.)

Slowly the reason for the extreme sensitivity and the seemingly erratic behavior of semiconductors began to dawn on physicists studying the problem. Small amounts of a foreign impurity do not alter the behavior of metals very much because there are so many free electrons anyhow. And small admixtures matter little in insulators because the electrons are not free to move. Semiconductors, however, have very few free electrons, so they are greatly affected by the addition of foreign atoms with extra electrons, which measurements of the Hall effect had shown were very mobile in these crystals. Research thus turned more and more toward understanding the role of foreign impurity atoms within semiconductors. The goal was to be able to change the conductivity of the substance at will.

The unusual phenomenon of increased conductivity at higher temperatures also became less puzzling. The warmer a crystal is, the more likely it is that an electron-donating foreign atom will not hold on to its additional electron but will donate it to the crystal as a whole. Conductivity increases when the crystal is heated because there are more free electrons, even though it is then slightly more difficult for these electrons to flow through the increasingly agitated atomic structure.

The potential of these materials was a source of delight and fascination to researchers. Yet even toward the end of World War II, semiconductors did not seem very promising for technology: the crystals were simply not chemically pure enough. Some German scientists believed that no systematic, pure substance could ever be a semiconductor, and they treated semiconductors like untouchables.

In the 1930s researchers, including Mott in England and Walter Schottky of the Siemens Company in Germany, had begun to delve further into the workings of Braun's crystal detector. When a needle tip contacts a crystal, current is rectified; it can pass in only one direction. Although vacuum tubes were being used as rectifiers, the crystal apparently offered a much simpler technical solution, if the action inside the crystal could be understood. Schottky believed that when the detector problem was solved, it could start a whole new technology. He thought that the answer would be found at the interface of the metal needle and the semiconductor surface. He pointed out that in the narrow zone at the interface, the semiconductor crystal is depleted of most of its free electrons. Electron depletion at an interface was already known from electrical batteries, such as those in automobiles. When an electrical voltage arises at such an interface, the current-carrying electrons must overcome it. Rectification can be explained with regard to positive or negative polarity of the voltage: one polarity raises the barrier to current flow, and the other lowers the barrier and eases flow.

Schottky's theory of the interface confirmed Braun's hesitant conclusion that the most important phenomena did not occur deep inside the crystal but just below the surface of the semiconducting mineral. Only a fraction of the crystal was active; the rest was merely a handle. Scientists could finally explain why every impurity and disturbance on the surface influenced the semiconductor's electrical properties. A layer of gas, with its charged atoms, or ions, can change the balance of the charges at a crystal surface and affect the current flow. A metal quickly equalizes such modifications by slightly displacing some of its many free electrons. Semiconductors do not possess nearly enough electrons to do this, so variations cannot be easily ironed out.

A New Research Age

By the end of World War II, the focus of major new research had changed, and the old German empire of scholars had been destroyed. The traditional form of physics research at secluded, small-town universities was dead. For scientists around the world, research was now greatly influenced by state military interests. During the war, heavily subsidized laboratories had proven their worth to everyone. Advancements in submarine and bomber technology had been worked out by hand-picked teams of specialists. Divergent interest groups were united skillfully, and new strategic weapons were developed rapidly. Huge sums of money poured into the new research, especially in the United States. The U.S. radar program cost an estimated two and a half billion dollars; the atom bomb undertaking, about two billion. Americans welcomed the new inventions and increased their respect for science. The country honored a new breed of hero, who had helped win the war and saved countless lives. The scientists who had fled from the Old World, especially Germany and Austria, helped America take the lead in physics and electronics.

Even the style of scientific teamwork changed. The government, once mistrusted, had proven itself a trustworthy employer. American politicians and military leaders had learned how to use science. Naturally, there had been tension and conflicts, especially during work on the atom bomb. Nevertheless, mutual respect was much more evident in the United States than in Germany or Japan. With government support and a strong economy behind them, American scientists turned their knowledge and experience to peacetime projects.

Harnessing the newly discovered atomic energy for peaceful uses was paramount. Radar, too, showed great potential for peacetime application and was soon adapted to weather forecasting and air traffic control. An additional challenge, not nearly as spectacular or emotional, was the search for electronic inventions that every household could use. Researchers finally understood crystals and semiconductors and hoped that they could gain a respectable position in the world of science. Soldiers had carried crystal receivers on their backs as part of their field telephones. They required little energy and storage space; perhaps they had future uses in radio and television technology. But the breakthrough with crystals occurred where it was least expected: in the research laboratories of a telephone company.

4

The Mother of Invention

. . . .

WHEN you make a phone call, you can neither see the path your message takes nor watch your voice carry to the person called. It seems almost miraculous that by dialing a number, you can reach one specific point out of hundreds of millions the world over. The key to the telephone network is its switches. Each call is sent to the right place because an enormous number of switches ensure that it is on the correct path. The switches have to be dependable and not use too much energy. The first telephones used mechanical switches. A weak control current was sent into a magnetic coil, which moved a metal reed that opened or closed the circuit to allow the main current to be switched on or off.

The giant American Telephone and Telegraph Company, founded by Alexander Graham Bell, foresaw by the end of World War II that its network would have to be greatly expanded to serve the country's growing needs. Mechanical switches would not be sufficient to handle this new load. New switches had to be developed to fulfill specific requirements: first, they had to have as few movable parts as possible; second, they had to use little energy; and finally, they had to be small. Simply miniaturizing the traditional mechanical switches would not solve the problem; the engineers pinned their hopes on the crystal; their vague idea was to use the electrons inside a solid as the only moving parts for a new type of switch.

"Leave the Beaten Track"

Fortunately, research and development were AT&T's strong points. In 1925, it had established a separate, independent research organization,

Bell Telephone Laboratories. The lab soon outgrew its first home on West Street in Manhattan and moved to Murray Hill, New Jersey. The establishment of a separate organization allowed the researchers more freedom to investigate as they saw best. Written on the foyer wall at Murray Hill are the words of Alexander Graham Bell: "Leave the beaten track occasionally and dive into the woods. You will be certain to find something that you have never seen before." This advice provided the industrial scientist with a coveted passport to independent research, which financiers and factory managers usually mistrust and dislike because it is expensive. The extent of research at AT&T was impressive; even in the 1940s, the company had more than 6,000 researchers and development engineers working on communications engineering—at a cost of several hundred million dollars.

Marvin Kelly, head of research at Bell Labs, circulated the edict that the mechanical relay had to be replaced. Researchers were to explore every technique and idea that might lead to new kinds of switches, amplifiers, and storage devices. At the center of their research was the crystal, the key to a new switch that would not require clumsy, separate components. One could use electron tubes as switches, but these awkward devices had to be heated white-hot and could never be produced in quantities sufficient for the telephone system. The goal was to replace these tubes while retaining their function.

In 1929 Walter Brattain joined Bell Laboratories. After an apprenticeship with vacuum tubes, Brattain began to try to reproduce their function in crystals. He was soon joined by William Shockley, a young, ambitious theoretical physicist who had worked in solid-state research. Shockley was convinced that there were two possible research routes for crystals: they could try to construct a kind of tube in the crystal, as Pohl had once tried, but with more suitable materials, or they could attempt to regulate the crystal's conductivity with an external control.

Other researchers had given up on the first possibility, but Shockley and Brattain tried it again in 1939. A tube consists of three main parts. Electrons issue from the heated cathode and stream to the anode. Between them is a grid for regulating current. Using fine wires, the researchers tried to build this kind of control grid into a crystal of cuprous oxide. This compound of copper and oxygen had proven itself useful in photoelectric cells and as a rectifier, although the mechanisms were not yet understood. But the experiments were complete failures. No change in current appeared; the crystal simply refused to react. Electrical conductivity seemed to be determined by other, uncontrolled processes, probably by contaminants in the crystal. The Bell researchers were greatly disappointed. As scientists, though, they were used to setbacks, especially in the pursuit of new frontiers. Fortunately, Bell's team leaders encouraged and motivated their workers. This attitude has always been Bell Laboratories' greatest asset.

Shockley and his colleagues turned to theoretical investigations of the detector problem that Schottky had outlined. Shockley's topic was the rectification of electric currents in the surface boundary layers, now called Schottky barriers. Shockley wanted to find out if some external influences in the surface layers could control and amplify current. On December 29, 1939, he wrote down this basic idea, which much later led to the Schottky field effect transistor. In the meantime Shockley was summoned to do research on German submarines for the U.S. military.

Marvin Kelly did not forget Shockley. Toward the end of the war, Kelly tried to persuade him to return to Bell Labs, and after Japan's surrender, Shockley did return. He was joined by another experienced crystal researcher, John Bardeen, who had studied theoretical solid-state physics in Wisconsin and at Princeton. He had made some major contributions to this field, and it was surprising—at least to Europeans—that a young theoretician dedicated to academic research would join an industrial team. Shockley, Bardeen, and Brattain decided to tackle the semiconductor challenge together. As a team, their backgrounds included theoretical and experimental physics, chemistry, crystallography, metallurgy, electronics, and precision mechanics. The Bell management helped to iron out conflicts and ensure that competition between team members remained reasonable.

Eureka!

The intense concentration on radar research during the war had narrowed the range of practical experience with crystals. In the postwar period cuprous oxide and selenium took a back seat, although sizable quantities were still being processed in the factories. The research team at Bell resisted the pressure to continue developing conventional switching devices and reducing the costs. They insisted that the best scientific prospects lay in the semiconductors germanium and silicon. Germanium had already passed the test as a detector crystal, and toward the end of the war silicon joined the ranks. Because both substances are elements consisting of only one sort of atom, they do not present as many problems as more complex crystals. In addition, they were recognized as true semiconductors, not contaminated metals. In their spatial arrangement, they follow the lattice pattern of diamond.

In the diamond, as in germanium and silicon, each atom is surrounded by four neighbor atoms, establishing the symmetrical tetrahedron, the three-sided pyramid. The structure of organic compounds is determined by the tetrahedral bonding of carbon, and now germanium and silicon opened up a new, inorganic world based on this structure. Diamond crystals sparkle when refracting light. They are also brittle and, unlike metals, do not pliantly adapt themselves to external forces; instead, they splinter or shatter. The strict spatial alignment of the bond of every atom to its four neighbors prevents the atoms from easily displacing each other. A metal, in contrast, does not have this spatial rigidity of the bonds. The electrons in metal cohere in a formless mass. A metal is thus softer and yields under a load; it can be shaped. Human civilization and culture as they have developed over the millennia would be unthinkable without metal's workability. This is a macroscopic trait, a characteristic of the entire piece of a material. Research on silicon and germanium spotlight a microscopic feature; man learned to use and command the nuances of these atoms and their bonds.

Elements in the fourth, central column of the periodic table have atoms with four electrons in the outermost orbit, which create four mutual bonds, each sharing two electrons, between neighboring atoms. Every silicon or germanium atom must use its four outer electrons to fulfill the bonding obligation and produce the tetrahedral structure. No atom will donate an electron to the crystal as a whole. Germanium and silicon in their pure chemical states are therefore not conductors. In contrast, every atom in a metal may deliver an electron to the total structure, making it an excellent conductor. Diamond is the ideal insulator, as it has absolutely no free electrons for conducting current. Silicon and germanium did not seem to fit this category, however. The seemingly obvious difference between the transparent diamond and grayish, opaque germanium and silicon had led to the earlier association of the two semiconductors with the metals tin and lead rather than with diamond, even though all are in the same column of the periodic table.

The quantum theory resolved the problem by explaining how energy releases one of the electrons in the tetrahedral bond. In a diamond the energy required to release an electron is so great that the photons of visible light, which have little energy, cannot do so. Light passes unabsorbed through the diamond crystal, making it transparent. In silicon and germanium, much less energy is needed to free electrons from their tetrahedral bonds. The energy of visible light is sufficient to release them, so both semiconductors are opaque to daylight. Light cannot pass through the crystals; instead, the light is consumed in freeing an electron from its bond, which leaves a hole in the bonding arrangement. Both the electron and the hole are then free to wander through the crystal. Thus it becomes clear why electrical conduction increases sharply when a semiconductor is illuminated. Exposure meters in cameras had used this phenomenon long before quantum theory could explain it.

What now seems so simple and logical demanded years of systematic work to decipher. The new class of materials finally received the long-overdue respect and recognition it merited. Basic research also uncovered the secrets of its electrical conductivity. If one replaces a silicon atom in the lattice with a phosphorus atom, one additional electron is donated to the crystal because phosphorus has five outer electrons, one more than silicon. Four are needed to form bonds, but the fifth is free to move and carry current, leaving the donor atom with a positive charge. Phosphorus, arsenic, and antimony, all from column five of the periodic table, are thus the donor elements, increasing the number of electrons in a crystal and enabling it to carry current. The magnitude of conductivity can be regulated by the amount of donor material added; very small admixtures produce great changes in the crystal, as had been noted before but not understood. Now the electrical behavior of a crystal of germanium or silicon could be changed over a wide range and very precisely—a great advance.

Even more important, by using the anomalous Hall effect, conduction was no longer restricted to negative electrons; not just donating but also accepting atoms could be built into a semiconductor. When a boron or gallium atom, both from column three of the periodic table and with three outer electrons, replaces a silicon host atom, one electron is missing from the bonds with neighboring atoms. An electron deficiency, or "hole," is thus created. To complete the fourfold bond, the boron atom accepts an electron from the crystal, introducing a mobile hole into the structure. In much the same way as an air bubble moves upward in a glass of water, the hole moves through the crystal, always in the opposite direction of the electrons. When current is applied, the hole moves just as if it were a particle with a positive charge. A simple analogy to cars being moved in a garage (see Fig. 3) has helped many people understand this baffling idea.

The concept of holes—missing electrons—proved extraordinarily important for microelectronics. William Shockley even named his famous early textbook *Electrons and Holes in Semiconductors*. The electron particle and the missing particle, the hole, demonstrate a very important principle of modern physics: the particle/antiparticle pair. Particle and antiparticle are opposites that can annihilate each other. When an electron meets a hole, both disappear, leaving an intact crystal bond. And when a bond is broken by an external force, such as light of sufficient energy shining on a solar cell, an electron/hole pair is generated. The generation and annihilation of particle and antiparticle also take place, at much much higher energies, in the particle accelerator facilities where the elementary components of matter are studied. Every particle must have its antiparticle—this general principle illustrates the basic symmetry of the physical world.

The Bell research team had justifiable hopes of finding a new way to regulate current and to replace mechanical relays with crystal switches. The most logical and straightforward way seemed to be by using the variability of current conduction in crystals and finding a simple way to modify the current flow by external means. Consider two plates, one metal, one a semiconductor, touching each other. If a positive charge is put on the metal plate, the semiconductor will acquire a negative charge to compensate for the positive charge of the metal. In this way an electrical field is created. The semiconductor, with its newly acquired negative charge, ought to be more conductive. By adding charge on the metal plate, it should be possible to change the strength of the current through the semiconductor. The same action that takes place in a mechanical relay should be possible, and far simpler, with this arrangement.

In 1926 the German physicist Julius Lilienfeld had already applied for a patent for this concept, the theoretical forerunner of today's field-effect transistor. Lilienfeld not only was the father of the field-effect transistor, a key concept of modern microelectronics, he started the brain drain from European to American laboratories. Right after applying for the patent, he left his German university to emigrate to the United States. His dream was to turn his ideas into a rewarding business, but his hopes were dashed when nature proved to be more com-

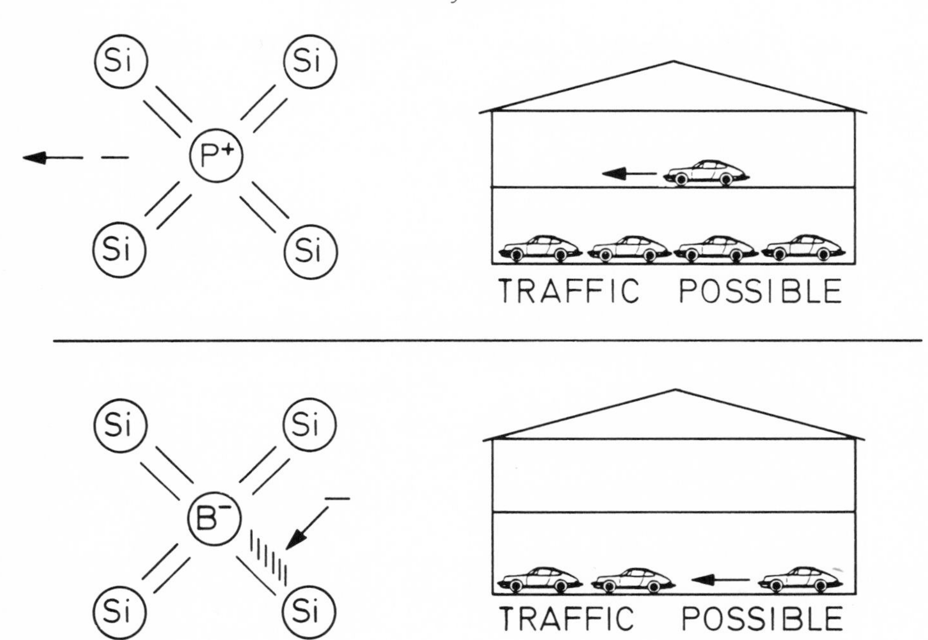

Figure 3. Donor and acceptor atoms in the silicon lattice allow the crystal to conduct electrical current. Upper left: a phosphorus (P) donor atom, with five outer electrons, can fulfill the bonding obligation to its silicon (Si) neighbors, indicated by the four shared electron pairs, and donate one free electron to generate current (electron with arrow). By donating the electron, the phosphorus atom becomes positive (P^+) and remains fixed in the lattice. Upper right: Shockley's analogy of a parking garage with all sites on the lower deck occupied, just as all the bonding sites in the crystal are filled. On the upper deck an additional car can move freely and allow traffic to flow. Lower left: a boron (B) acceptor atom can supply only three electrons to the bonds with its neighbors. A "hole" (shaded area) remains in the array of bonding electrons. A free electron may jump into this hole, thus generating current flow. After accepting the electron, the boron atom is negatively charged (B^-). Lower right: in Shockley's garage analogy, a vacant parking site allows cars on the lower deck to move. It is much simpler to describe such motion as being that of the vacant site, or hole. The hole moves in a direction opposite that of the electrons, thus behaving as if it were a positively charged particle.

plicated than the simple drawings in his patent application. The Bell Labs team later experienced the same obstacles.

Although the concept of a semiconductor switch appeared simple, nature continually frustrated actualization. The scientists tried every variation of the drawings in the patent specifications. The procedures seemed so convincing on paper! Conductivity was improved by changing the number of electrons in the semiconductor, but there was still no change in current flow. The Bell team used every investigative method possible, ranging from rigid, theoretical predictions to obscure experiments reminiscent of black magic.

They did discover that when they brought a semiconductor sample into contact with a liquid, they could sometimes detect an improvement. Brattain, the pragmatist, especially pushed in this direction. Shockley, the purist and theoretician, felt the experiments were drifting too far away from the primary goals. John Bardeen took the experiments seriously and hypothesized that something on the crystal's surface prevented the desired change in the number of electrons inside the semiconductor crystal. The researchers knew that the field effect altered the charge number in the semiconductor, although it could not be observed in flowing current. The additional charges could not be induced to join the current. Apparently, there were traps which caught the electrons, not inside the crystal but on the surface. The crystal's surface characteristics had once again caused a predicament.

Nobody had expected the atoms on the surface of the crystal to have such an impact on the process as a whole. However, physicists and chemical engineers learned that these minorities did determine the behavior of semiconductors. They had the example of donors and acceptors as a guide, so it seemed logical to search for the root of the problem in the surface atoms. All of those depressing, frustrating experiments gave birth to a productive branch of basic research: surface physics.

On the surface of the crystal, the solid's most important property, symmetry, is abruptly destroyed. The continual, rigidly determined, systematic arrangement of its atoms is suddenly halted. In physics, any break in symmetry, any disruption of the system, brings on drastic reactions. The study of symmetries and their disturbances is an important aspect of the natural sciences. The breaking of symmetry affects quarks, magnetic fields, stellar structure, even ice. In the semiconductor the symmetry of the crystal lattice exerts great influence on the arrangement of the electrons and their energy states. At the surface the break in symmetry changes the energy states. The researchers found that localized surface states develop, which finally showed that a breakdown of symmetry is the cause of those problematic surface traps that immobilize electrons.

The next step was to find a way to influence the surface. The researchers looked to chemistry for the answer. What if one of the neglected surface atoms of the germanium crystal were provided with a new chemical partner? Some liquids seemed to provide a means to do this. The team now needed measuring probes that would provide precise data on fields, charges, currents, and voltages, to study the exact conditions on location. It seemed natural to turn back to Braun's point contact. This time, however, numerical predictions were possible.

The scientists used a drop of liquid as a contact in their new experiments with germanium. Peering through a microscope, they used a fine needle tip to map the surface regions of the crystal. Walter Brattain was an especially valuable team member during these tests. He immediately recognized any deviations from the norm, so important to experiments in physics. Brattain noticed that at a certain polarity of the needle tip, resistance changed at the interface of the germanium crystal and the drop of liquid.

The team picked up speed in those last days of 1947. It appeared that the little needle tip could shoot additional holes into the germanium crystal, which would change its resistance and therefore the potential. On December 16 the researchers experimented by using two needles without any liquid and achieved the same results. They then painstakingly prepared a germanium crystal surface. Tiny dots of gold were evaporated in a vacuum to act as the external contacts. Under a microscope the needle was cautiously touched to the contacts, which supplied external current. When one of the two contacts was under voltage, the current flowing into the crystal via the other contact grew stronger. The researchers' excitement was justified, for they had found a true crystal amplifier, even if still a very weak one.

The team had proven that a surface contact could have an effect deep inside the crystal, even though the contacts were only about one tenth of a millimeter apart. Compared to the space between electrons inside the crystal, this distance is enormous. An electron can travel only a very short distance without being impeded. The big surprise was that the charge-carrying electrons or holes, injected by a contact, did not weaken as they traveled. Electrons can cover great distances in a vacuum tube, for they seldom collide with one of the few remaining air molecules there. The crystal, on the other hand, is tightly packed, full of traps and obstacles.

Bardeen and Brattain calculated that they could amplify electric power if they could get the contacts to within two thousandths of an inch without touching each other. In today's microelectronics this distance is huge. At that time, however, scientists had to use all their skill to attain such minute dimensions. Brattain and his associates evaporated gold onto the edges of a small plastic triangle. Using a razor blade, they cautiously separated the gold overlay on the tip of the triangle, producing two independent leads. They next touched the triangle to the germanium crystal. The experiment succeeded on the first try, and the first transistor was born!

Groups meeting casually in the corridors of Bell Labs talked of nothing but the major breakthrough. Colleagues from the office next door were called in for their objective comments. In all the commotion, the team was afraid of making a mistake. Every day they meticulously entered their findings and drawings in the patent notebook and speculated about further developments. One colleague always witnessed and signed the patent papers with "read and understood."

Because it was shortly before Christmas, everyone worked in a frenzy to present the directors with a finished product before the long holiday weekend. Their intense work paid off. After the successful experiment with the plastic triangle, they developed a new and simpler technique. They brought close together two extremely fine metal tips sharpened to a point in an etching solution, then pressed them onto the germanium crystal. The point-contact transistor proved quite powerful with this setup.

The team next looked for current amplification by means of the new transistor. Many seemingly good inventions in electronic components failed in practice because they could not amplify electric current. In an electronic circuit, as impulses are transported back and forth, each step weakens the signal, since resistance in the leads consumes energy. An amplifying element, however, can draw on an energy storage unit and regenerate the signal. This ability to regenerate, already a significant factor in the first transistor, later became absolutely essential for integrated circuits. Only an amplifying component can produce oscillations; the amplified signal must be fed back to excite the up-and-down motion.

The researchers hastily built a circuit to convince everybody that their new germanium device was indeed useful for amplifying small electrical signals. This kind of proof is what patent lawyers call the "reduction to practice," a demonstration beyond a mere diagram on paper. The circuit was of the type used in a loudspeaker system to amplify speech. On the afternoon of December 23 the Bell Labs directors were invited to witness the demonstration. Brattain's notebook records: "This circuit was actually spoken over and by switching the device in and out, a distinct gain in speech level could be heard and seen on the scope presentation with no noticeable change in quality."

Nobody can now remember the first words that were amplified by a piece of crystal. But the day—December 23, 1947—is legally fixed as the exact date of the invention of the transistor. As Shockley reminisced, "In any event, it was an exciting start for a four-day Christmas weekend. The anniversary of the invention coincides with the warm spirit of the Christmas season."

In official photos, staged at a later date, Bardeen and Brattain look on as Shockley peers through the microscope at the transistor. The microscope had special significance. Earlier inventors stood next to large, tangible objects, such as steam engines, cars, and motors. For the first time, a great new invention was drastically smaller than its inventors.

Researchers spent several weeks thinking up a name for the new device. John Pierce suggested "transistor." The suffix *-istor* commonly designated the family of these electronic elements. The prefix *trans-* implied that conduction traveled *through* a piece of crystal. The two needle point contacts were also christened. The point contact that injected the charge carriers was named the "emitter"; the other one acted as, and was accordingly termed, the "collector."

Inside the Crystal

The crystal had passed one of its toughest tests, a principle had been demonstrated, a possibility shown. The big surprise was that the injected carriers of electric charge, that is, electrons or holes, were effective over distances of the order of several hundredths of an inch. Such dimensions seem small on a human scale, but they are really surprisingly long when compared with the distance an electron travels before being deflected. The crystal must be perfect to keep the injected electrons alive and continuing on their journey.

The motion of atoms and electrons in a crystal ensures electrical equilibrium, but it takes a long time for extra electrons or holes to dissipate. In metals there is absolutely no possibility of maintaining one electron too few or too many; the electron soup immediately levels out the imbalance. In a semiconductor, though, electrons can survive for almost a thousandth of a second, which again seems incredibly short on a human scale but is very long when measured by the time between an electron's hitting an obstacle within the crystal and changing its direction—something less than a trillionth of a second. A lifetime of a thousandth of a second is long, because electrons crisscross through the crystal lattice at speeds of ten thousand yards per second; therefore they can indeed travel long distances to get to where they are needed, as at the collector.

Scientists could finally explain some of the difficulties and failures that had plagued their predecessors. They had ascertained that these long lifespans and paths of the electrons were possible only in perfectly clean semiconductor crystals. Every disorder, every missing or dislodged atom, every contaminant atom endangers the life of the added electron. Crystal electronics demands materials research. The next step, therefore, was to measure and identify all possible disorders and disturbances and learn how to avoid them. What was needed was the ideal crystal.

The point-contact transistor was turned over to the development and production engineers. To achieve mechanical stability, they placed the germanium in a sturdy tube. Two legs protruded from the top, and one from the bottom. But the engineers found this flimsy arrangement frustrating. Their complaints were not unfounded. During the war one cat's whisker had done just an adequate job in radar, and now they needed two of those rather unreliable leads. It was not just the two whisker tips that posed problems, there was also the difficulty of maintaining the exact space between them. And the ultrafine tips of the needles could barely transport a strong current. Given all these problems, the transistor made a poor showing; the whole configuration seemed jury-rigged. Besides, the original plan had been to house the whole thing inside the crystal.

Shockley was hurt. He had entertained bigger plans, which now had to take a back seat to Bardeen and Brattain's success. When (two years later!) the prestigious *Physical Review* published a detailed report on the transistor, Shockley's contribution was not acknowledged until almost the last line. The authors of the eighteen-page publication were J. Bardeen and W. H. Brattain. Although he was the team head, Shockley was not automatically entitled to be a coauthor of one of the team's scientific publications. But Shockley was an extremely ambitious and hard-working researcher and had poured himself into this project. He made no secret of his intense disappointment at being nearly forgotten. The company had to mollify its team with diplomats and arbiters.

Many years later, though, Shockley wrote how one could turn such a personal setback into an impetus. He emphasized that the researcher must learn from "creative failures," using gaps and mistakes to recheck scientific principles. Shockley always drove himself to go beyond conventional thinking. Nuclear physicist and Nobel laureate Enrico Fermi had once told him that good research was "based on the will to think." The effective researcher tries to continually rethink the traditional, the apparent, the tried and true. Shockley always castigated his co-workers when they made vague explanations. Demanding quantitative statements of the facts, he would not tolerate imprecision. Impatience, stubbornness, and irritation drove him on. His maxim was: search for the "simplest case."

Looking back on the germanium experiments, Shockley questioned why such a clumsy setup was needed to shoot holes into an electron-conducting crystal. Why not attach to the crystal a layer that had sufficient holes? Then an electric current could drive the holes out of this layer and into the neighboring portion of germanium. Perhaps the old-fashioned needles could be dispensed with.

The boundaries between layers in this kind of arrangement merit special mention. As in life, most of the important events in technology occur at junctions. Consider the living cell, whose protective but porous membrane wall permits the exchange of substances and energy with the surroundings. Just think how many complex chemical reactions depend on interactions between surfaces. Our lungs and circulatory system consist almost entirely of surfaces that permit a carefully monitored exchange. Boundaries and borders are raised or lowered by heat or electrical signals. The Bell Labs researchers, searching for just this kind of controllable border, finally found it in the *p-n* junction.

Germanium with electron-accepting impurities is called a *p*-type (positive) semiconductor. The positive holes, the antiparticles of the negative crystal electron, define this kind of acceptor-doped material. A semiconductor doped with electron-donating impurities is called an *n*-type (negative) semiconductor. In this case the electrons determine the semiconductor's behavior. Shockley postulated that something *must* be happening at the interface of *n*-type and *p*-type material. Where electrons and holes confront each other, a thin surface layer is left with no charge carriers. It is almost impossible for current to flow unless a positive voltage repels the holes and pushes the electrons across the barrier. A negative voltage applied to the *p*-zone does not generate any current. On the contrary, it draws even more holes away from the junction and increases the resistance. Ferdinand Braun's rectifier had finally been truly achieved: current flowed in only one direction; the barrier prevented flow in the other. And it all took place inside the crystal, at its internal interfaces.

Shockley's notebook gradually filled with increasingly precise formulations as to how a "proper" transistor should look. Around New Year's Eve of 1947, Shockley spent several days in a Chicago hotel room, filling pages and pages with possible setups. He still did not see how electrons and holes in the crystal might travel such great distances to reach a spot where they then paved the way for current to follow. The idea of using controlled injections of minority charge carriers, such as holes in an *n*-type region, came later. A brilliantly devised experiment eventually confirmed his ideas.

One late January evening, Shockley and a colleague were riding the Erie-Lackawanna train home to New Jersey from New York. Shockley spent half an hour clearly explaining, he thought, how electrons would have to pass through a thin layer to reach the other side. Just before they reached Summit, their destination, he postulated the necessary mathematical formulation. Shockley eagerly awaited his friend's enthusiastic reaction, questions, and suggestions, but he got no response at all. Not having understood the lengthy explanation, his coworker was unable to agree or disagree, proving once again that convincing a critical friend is the best test in physics research. Shockley was deeply disappointed and had to admit that his formulations were still too vague.

On April 7, 1949, after many treacherous detours and dead ends, the concept was finally proven to be functional. The team confirmed the theory under conditions that would stun today's semiconductor experts. Shockley wanted two *p-n* junctions, one to generate a flowing current, the other to act as a barrier. *P*-type germanium was melted and a drop of it was poured onto the *n*-type crystal, just as one pours hot pudding into a cold mold. Using a saw, an assistant cut the solidified blob to size. And indeed, the junction transistor worked! It certainly did not *look* good, but it obviated the need for those troublesome needle tips.

For the next two years Bell researchers dedicated themselves to refining and improving this junction transistor (Fig. 4). It is known as a "bipolar" transistor because it involves two polarities, positive holes and negative electrons. The junction between, with its changeable barrier height, is the most important part and lends the transistor its name.

As transistor research and development progressed, it became obvious that the quality of the crystal used was decisive. The materials were no longer created like hot pudding, which does not solidify into a sufficiently symmetrical structure. Instead, the crystal must be allowed to form slowly and carefully. Some researchers remembered the old experiments and procedures for making large single crystals, but the semiconductor experts were not enthusiastic about this seemingly too complicated method. They mistrusted such suggestions because they were "not invented here," a syndrome that often affects top researchers. But experiments made it plain that only a chemically pure, crystallographically perfect crystal would work in semiconductor electronics. Scientists soon perfected the technique of drawing large, high-quality single crystals out of molten germanium. The cylindrical crystals were then cut into wafers and further treated.

The fundamental physical laws governing control and amplification of current in a solid were at last understood. All of the major processes took place entirely inside the crystal; no longer did engineers have to

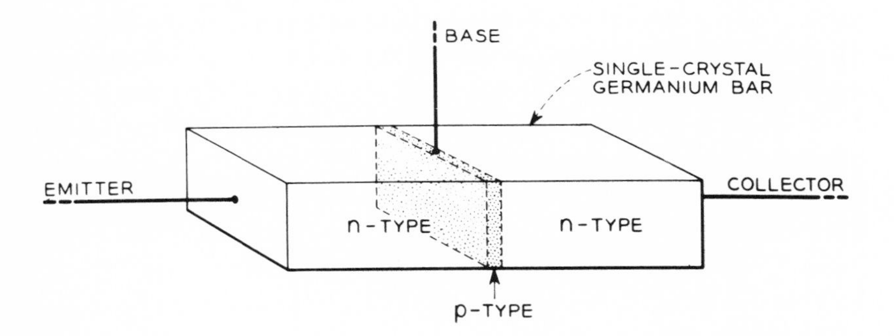

Figure 4. Diagram of a junction transistor. A fine wire is connected to the thin central section of the germanium bar. The two end portions of the bar have different electrical properties from the central section.

struggle with complicated external arrangements. The crystal's own structure, with some atoms donating and others accepting electrons, had finally made it functional.

Nobel Prizes for Industrial Research

In the fall of 1956 when the Swedish Academy of Science convened to decide on the winners of that year's Nobel Prizes, they gave the physics prize to Bardeen, Brattain, and Shockley for inventing the transistor and for semiconductor research. The awarding of a Nobel Prize to a team from an industrial research laboratory—Bell Telephone Laboratories—was extraordinary! University researchers were the usual recipients, mostly for pioneering work in atomic or nuclear physics. Yet it was not the first Nobel Prize for physics that a Bell researcher had won. In 1937 Clinton Joseph Davisson had been honored for discovering that electrons not only look like particles, they also stream in waves through thin metallic foils. His experiments had confirmed an important theoretical prediction and had proved vital for tube technology. And later, in 1977, Bell researcher Philip Anderson received a Nobel Prize for explaining the seemingly contradictory dual nature of electrons in crystals, which sometimes traveled freely and at other times remained at fixed locations.

Arno Allan Penzias and Robert Woodrow Wilson also won the Nobel Prize in 1978. They had tried to construct a highly sensitive antenna for satellite reception but were frustrated by an inexplicable source of noise, crackling static. In continued experiments they found that the culprit was not the antenna but radiation remaining from the Big Bang! Their breakthrough in understanding the origin of the universe was made possible because of the freedom given them at Bell Labs to conduct basic research. This tradition of top-notch basic research in the laboratories of American industrial firms led to striking successes in 1986 and 1987, when for two consecutive years researchers at the IBM laboratory in Rüschlikon, Switzerland, near Zurich, won four Nobel Prizes for physics.

In the *Journal of the American Academy of Arts and Sciences,* Harvard professor Harvey Brooks wrote that the Bell System was the best example of a highly integrated structure in a high-tech industry. The System is recognized throughout the world as a highly successful and innovative technical organization, and Bell Labs became a model for many large firms in Japan and the United States. A company that nearly monopolizes the entire communications network of a continent has tremendous financial resources that are independent of business fluctuations. In West Germany, the telephone system, part of the state-operated German Postal Services, is bursting at the seams with money, yet it stingily neglects basic research. Bell researchers had the advantage of being part of a state-regulated monopoly.

For AT&T it was a principle of business to invest in research and development and thus ensure the network's future. Bell Laboratories' management believed that the research it conducted should conform to rigorous quality requirements and contribute to the national welfare. Bell Laboratories also cooperated on important projects with the federal government, including the first telecommunications satellite, Early Bird. And the company was active in nuclear weapons research through its Sandia affiliate in New Mexico.

To Europeans, who were accustomed to a totally different research environment, American research organizations, and especially Bell Laboratories, offered many unusual organizational methods. The traditional European personnel policies regarding tariffs, salary increases commensurate with age, and civil servant status were not so rigid. American methods of recruiting the most promising young scientists came as a surprise. Bell Labs and other firms regularly sent small teams of experts to the country's leading universities, where they took note of the best professors and recruited top students. The teams offered financial support and arranged interviews long before prospective employees received their doctorates. The candidate had to give a presentation, often attended by Nobel laureates and well-known scientists, then be subjected to questioning, a far tougher test than a university doctoral exam! At Bell the managers then questioned not only the applicants but also their interviewers, to find out where they would most like to work.

Most European research organizations, especially state-run institutions, are bound by rigid salary scales and seniority clauses and are often headed by military or legal branches. Such organizations can suffer from their very size and type of leadership. Bell has managed to avoid a staid, static bureaucracy even though it has 15,000 employees. Through the years Bell has acted as a national elite university for the United States, educating new solid-state researchers. America's lead in the new semiconductor industry can be traced to the superiority of its industrial laboratories.

5

Transistors Come of Age

AT the beginning, Bell Laboratories kept the transistor, its great new discovery, a secret. When a young assistant professor from Purdue innocently told Brattain about his own experiments in semiconductors and asked him for advice, Brattain had to act as if he knew little about the subject. Six months later Bell Labs finally revealed the invention to the public. Instead of an outburst of enthusiasm, the public showed almost no interest. On July 1, 1948, the *New York Times* mentioned the transistor on page 46, in its regular column, "News of Radio." Listed first were the really important announcements of upcoming weekend entertainment programs: Mel Torme would be appearing instead of the popular Dinah Shore, Harry James would be playing Tuesday evening on NBC. The column continued with news about new radio shows and their sponsors, and at the very end the transistor was finally mentioned.

This was followed by a dry description of the technical specifications, probably copied from Bell's press releases. In its report the *New York Herald Tribune* warned: "However, aside from the fact that a transistor radio works instantly without waiting to warm up, company experts agreed that the spectacular aspects of the device are more technical than popular." Naturally, the public was not impressed, and the scientific journals were not a bit excited. For the first year of its existence the transistor remained a kind of laboratory curiosity. British researchers even considered the transistor a clever publicity stunt by Bell! The more novel an invention is, it seems, the stronger the resistance to it. With no ready consumer market, this reaction is not really surprising. In this case, though, the initial reaction was disappointing, because the transistor had started out on a much more favorable footing than many products that have sprung from scientific laboratories.

A Time of Transition

It was not just the market that showed resistance. Legions of engineers viewed this new technology as an undesirable intruder into their safe and known world. Because semiconducting materials react oddly to even a few atoms of a foreign substance, research labs and factories had to develop a phobia about contamination surpassing even that of operating rooms and pharmaceutical factories. If a worker is wearing gold jewelry while treating a germanium chip, the results might be ruined, because copper in the gold alloy could enter the germanium and interrupt the electrons' paths to the collecting contact. The crystal forced research labs and factories to work under sterile, exacting conditions.

In the new semiconductor technology, small was beautiful. A ten-thousandth of an inch, the typical thickness of the base layer between the emitter and the collector, became an ordinary measurement. This zone had to be so minute that electrons or holes could pass through it quickly and thus rapidly track the high-frequency electrical waves of radio and television.

Laboratory researchers continually refined their mastery of the tiny structures in the germanium crystal. Its physical nature became increasingly well understood, and the information more exact. Shockley wrote a textbook, *Electrons and Holes in Semiconductors*, which was the Bible of the field for many years. Scientists saw the dawning of a new age, but engineers were much less inclined to such poetic views, and carefully calculating businessmen remained unimpressed.

At least theoretically, the transistor had shown that it had considerable advantages over vacuum tubes, so its market was seemingly ready-made. Radios, televisions, amplifiers, transmitters, and other products using tubes needed only slight modifications to accommodate transistors, which were much smaller and, with mass production, less costly. They saved money by consuming less energy; wire filaments no longer needed to be heated, which made radios far more convenient to use.

By this time the amazing developments that had taken place in electron tube technology had given way to complacency. Feeling secure from any competition, tube technicians were not making any noteworthy improvements. But when the transistor suddenly came along, those working with tubes panicked and immediately began making tubes more compact, shrinking them to the size of an acorn. Rechristened with the modern name "Nuvistor," the tube promised the benefits of high technology to a new generation. But the old guard was roused only briefly before its final demise.

Inventions that have military uses, such as radar and related developments during World War II, often are able to enter the market readily. But the transistor, a peacetime invention, continued to encounter consumer resistance. It was still too expensive and unreliable; the familiar, available, proven, widely accepted vacuum tube held its own for a long time. In addition, the transistor had a bad name because of its initial defects. The old point-contact transistor, with its two cat's whiskers, had been much too problematic for practical-minded radio manufacturers and suspicious customers. The new generation of *p-n* transistors, whose operations took place inside the crystal, though promising, needed a long time to gain acceptance.

In the early transistor years Bell Labs harvested scientific fame but also received heavy doses of skepticism. For this reason the first use of transistors was not in radio or the telephone system. The hearing aid was the first transistor-run device. Alexander Graham Bell had experienced the tragedy of hearing loss in his own family and had given a great deal of time and money to research concerning deafness. In memory of its founder, the Bell System did not charge any license fee for transistors to be used in hearing aids. In the first years of its public life, the transistor was almost always associated with the hearing aid, which could now be worn invisibly. People who were hard of hearing no longer had to lug around a large battery or remain near an outlet. One of the first licensees for the transistor for hearing aids was the Raytheon Company. Only eighteen months later, more than a dozen new businesses manufactured these compact, convenient, energy-saving devices. Even with the batteries, they weighed less than three and a half ounces. Originally more expensive than hearing devices using the Nuvistor, the new aids gradually became less costly and required fewer expensive batteries.

Vacuum tubes, which cost just over fifty cents each, still dominated all the other important branches of electronics. But in old ramshackle buildings in East Coast cities, the new electronics firms attempted to drive down transistor prices and break down the use of tubes by the established manufacturers. The breakthrough in hearing aids, universally acclaimed, was followed by the invention of the medical pacemaker, which would have been unthinkable without the microelectronics of the transistor. The crystal devices became increasingly smaller and consumed less and less energy.

Bit by bit the scope of the transistor business grew. In 1952 only eight firms were processing germanium into transistors; in 1953 there were fifteen. The traditional electronics companies now recognized that they too had to make provisions for this crystalline replacement for vacuum tubes; new departments were created in the old tube factories. The transistor took a giant step forward when the recently founded Texas Instruments Company marketed the first portable transistor radio. The new product became an immediate hit and soon invited many competitors. Nations in the Far East in particular saw a chance for new exports and began producing cheap transistor radios. The germanium transistor was capable of picking up at least one AM transmitter. The loudspeaker and reception components required so little power that a single set of batteries enabled people to listen to music anywhere, any time. The new radio "toy" became so popular that it was soon dubbed the "transistor," which actually refers only to the germanium chip housed in the receiver.

As a result of World War II, Germany lost many of her assigned frequency bands for radio transmission in the Copenhagen Accord. Europe's dense population resulted in crowded radio waves, and in many parts of West Germany radio reception suffered interference. Necessity was the mother of invention: the ultra-short-wave range was developed to cope with these conditions. Much higher frequencies were needed, which made the range of the radio waves short. Improved switches and electron tubes were needed to process the more rapidly oscillating waves. Using the principle of frequency modulation (FM) of radio waves instead of the earlier amplitude modulation (AM), the new radios provided highly improved reception, but they were more expensive. To capture the full sound quality of a broadcast, large loudspeakers, housed in costly wood and metal trappings, were the order of the day. The radio became the center of attraction in every German living room. In comparison, the portable transistor radio seemed frivolous. It also seemed rude and unmannerly to carry music about everywhere, disturbing others. Advertising was rare on German state-run radio stations; industry had not yet grabbed the chance to advertise its goods and services.

In the United States and Japan, however, the transistor radio gave quite a push to technological development. The germanium transistor was no longer left only to the theoretical researchers; whole teams of developmental engineers were formed. Trained personnel became scarce. Only a few universities offered courses in the science of solid-state semiconductors, and such departments grew slowly. A top university was happy to hire someone for its faculty who understood the electronics industry, even if he was retired. One of the most serious bottlenecks in the advancement of technology was that university programs, as well as the training of technicians and laboratory workers, lagged behind, and this is still true today. Research tends to move so quickly that undergraduate teaching has a difficult time catching up with the newest developments.

In electronics the cost of developing a prototype to the point where it can be produced is about ten times higher than the cost of researching it. To make matters worse, it costs about ten times more to equip an efficient production line than it does to develop the product! Despite the significance of initial research, its cost is not an important factor.

Americans change fields more readily than Europeans, so it was not surprising that many different professions were soon represented in the new semiconductor field. America's nuclear bomb programs had trained and educated many nuclear physicists. After Japan's surrender, these scientists found that few job opportunities awaited them except at the universities. Many of these young physicists became semiconductor researchers. According to Harvey Brooks, dean of engineering at Harvard, nuclear physicists were by far the largest group of scientists to turn to semiconductor engineering. Most of them had no real background in the physics and technology of solid states and crystals. Along with the physicists were many biologists, who, finding few job possibilities in their own field, increasingly turned to the new solid-state research teams.

The U.S. military showed little interest at first in the technology of the transistor. Invented in a telephone research lab, it was certainly not a military project. Bell Labs, although always loyal to the government, had refused to undertake military research projects until 1949, when it became unavoidable. Bell treated the military professionally, just as it treated every other customer. Bell reported on all its new projects, but did not succeed in interesting the civil-servant military technicians in the transistor, which was considered merely an interesting variation of the tube.

In 1952, one year after Shockley's successful demonstration of the new *p-n-p* transistor, the Department of Defense established a subcommittee on semiconductor components, but it was still subordinate to the commission on electron tubes. Suddenly there was great interest in portable instruments. Every American child knew about Dick Tracy's watch with its built-in radio and telephone, which now was not just a science-fiction fantasy. The military wanted such a device for soldiers to use in the field; the price to develop it was irrelevant. The armed forces bought up almost all of the 90,000 point-contact transistors Western Electric manufactured in 1952, but the unreliable set-up with those two points caused a great many headaches.

A military report from 1955 noted with disappointment that the transistor had proven unreliable. In only twenty-five of one hundred units did the electrical properties remain within permissible limits. All the others changed their properties or failed altogether. Even these values deviated from those in the original data. The report described the unreliability and uncertainties of the transistor as revealed in tests: the measuring indicator would twitch back and forth, and drastic changes would occur when the transistor was connected to a military instrument. The U.S. Air Force estimated that in 1952 it had no less than a 40 percent failure rate with germanium transistors. It was impossible to equip soldiers with such undependable instruments. In succeeding years the military tried to make semiconducting devices more reliable and sturdier, regardless of costs and market aspects. Because of the money being poured into research and development, the transistor began to gain on the tube.

The American Telephone and Telegraph Company invested heavily in continued research, particularly in technical development. There were many areas where the transistor could be used, even if the mechanical switching relays could not be immediately replaced. Scientists registered patents for every imaginable component variation. This growing wealth of information was not the sole property of AT&T; the government had put some restrictions on the concern.

Although the telephone, telegraph, and radio companies had started as private concerns, Federal Communications Commission regulations turned them into half-state, half-private corporations. Legal authorities decided that the communications industries should be treated like the railroads, with certain obligations to the public. Communications companies had to publish their charges and dividends and get government permission for mergers and acquisitions. But at the same time the state guaranteed a system of markets and regions. This suited the technicians perfectly, for only a standardized network with clear-cut norms can ensure regulated traffic. All local telephone affiliates of AT&T had to buy their equipment from Western Electric, guaranteeing the compatibility of network installations. Technically powerful Western Electric, however, was not permitted to enter other markets; it could not produce radios or hearing aids.

What happened to the great store of knowledge Bell researchers had garnered? The Western Electric Company was the legal holder of all the patents acquired by Bell researchers. Bell and Western Electric shared all new information. A deliberately open attitude toward information sharing became a standard Bell policy. In 1952 anyone interested could take part in a workshop at Bell Laboratories and learn everything he wanted to know about the new point-contact transistor.

At a Bell Labs symposium held in April 1952, the state of development and the technical specifications of the *p-n* junction transistor were disclosed. The cost of getting this information, which had been gathered over years of intensive work, was ridiculously low, 25,000 dollars. And that price could be deducted from future license fees if the customer decided to manufacture transistors. Thirty-five potential customers came to the symposium. Large established electronics firms sent their representatives, but there were also some brave newcomers. A few years later three dozen firms in the United States and nine partners abroad were licensed to manufacture germanium transistors.

Bell reaped twofold profits from this generous act. First of all, the firm proved to the FCC that a giant company, one that dominated the research, was willing to make concessions, thus justifying its monopoly of the telephone industry. Second, Bell's new licensees brought fresh ideas for new products, accelerated the pace of technological advances, and pressed for greater reliability and reduction in the cost of each transistor unit.

The technological advances were less spectacular and scientific but more practical than before. General Electric felt compelled to enter the semiconductor field and developed a simple method for manufacturing the complicated layer structure of the emitter, basis, and collector. The GE scientists found it surprisingly easy to alloy indium into the germanium chip, which enabled them to produce the special electrical properties of each zone. GE's alloy process was easier and cheaper than the rather awkward process Bell had used. The big East Coast electronics companies soon incorporated the GE method, and Philco developed sophisticated techniques in this area. Using a jet of etching solution, scientists there were able to precisely reduce the thickness of the germanium crystal. This resulted in transistors that could be used for very high frequencies. They next placed the little ball of indium on the spot that had been thinned and alloyed it into the germanium. Their process was costly, but it improved the product yield and was particularly important for the European partners. In Europe germanium crystals had to be extremely thin in order to satisfy the requirements of manufacturers of FM radios operating on ultrashort frequencies.

American industry and universities strengthened and accelerated the development of these new technologies. Nevertheless, the lion's share of research remained at Bell and Western Electric. (Western Electric, the equipment-manufacturing part of AT&T, having paid for much of the research at Bell Labs, then developed the ideas further into usable components of the AT&T network.) The telephone company's great technological advantage and market dominance became increasingly obvious to the FCC; and in 1956 Bell was forced to renounce major rights on all its transistor patents or run the risk of being split up into individual, competing companies. That was the fate that had been dealt to the Rockefellers' Standard Oil Company.

AT&T chose to retain its monopoly and make its technical information cheaper and more accessible to all interested parties. The firm could continue to research and develop the semiconductor. But the patent consent decree did affect the further development of microelectronics in the United States and proved that the American government understood the political and economic significance of this growing technology. Foreign countries also profited from the decree. Japan, for one, jumped at the chance to use this wealth of scientific and technical information.

Bell Laboratories evolved into a kind of super school for the new technologies. Former researchers fanned out into all the corners of the United States. One firm in Dallas, Texas, immediately recognized the great opportunity. The pioneers searching for oil had learned to adapt highly sensitive seismographs, normally used for measuring earthquakes, to find oil. New methods of chemical analysis then could determine whether a potential oil well would pay off. A company named Geophysical Services Incorporated proudly renamed itself Texas Instruments and, after the symposium of April 1952, obtained semiconductor licenses from Bell.

For a number of years before the Japanese onslaught, Texas Instruments was by far the largest semiconductor company. It kept its eye out for the best experts and won over Gordon Teal from Bell Labs, who had lovingly raised the single germanium crystal to high levels of perfection and dependability. The Texas Instruments lab eventually developed into the world's largest semiconductor manufacturing plant. In this way, a small concern on the North Central Expressway in Dallas in time became the first to have an annual turnover of more than a billion dollars just from semiconductor components.

Calculating Machines and Computers

The transistor was given a big push with the development of electronic computers. One of the first tasks for the crystal was to assist in the performance of mathematical functions and the development of switching and storage components for computers. Components had to be small and use little material and energy; the information paths had to be short and problem-free.

In the seventeenth century the German philosopher and mathematician Gottfried Wilhelm Leibniz had wondered whether the drudgery of arithmetic tasks could be performed by machines. Leibniz, a sincere believer in the immanent harmony of nature, believed that for humans to grasp this harmony it was necessary to have complete, quantitative, and correct descriptions of natural phenomena. Mathematics, which began to emerge strongly during his time, seemed to provide for consistent description by figures and curves. He thought that by having a universal mathematical language with a formal system of expression, the study of nature could be free of the pitfalls of national languages. The philosophy of mechanism, which reduces all phenomena to simple mechanical processes, led Leibniz and Newton to develop differential calculus. With this new branch, mathematics was no longer limited to static accounting and balances; it could describe change with formal accuracy and relate cause and effect.

Differential equations enabled scientists to correctly predict the motions of the stars as well as trajectories and changes in populations. One of Leibniz's most important goals was to develop a universal calculating machine using precisely formulated rules that would derive results without any danger of subjectivity. He and his contemporaries also hoped that this kind of machine would help heal the schisms between the various Christian denominations and between Christian ethics and the philosophy of mechanism. He wanted to discard the abacus, the calculating instrument of Asian merchants, and replace it with calculating machines that would rescue humanity from fateful coincidence and bring about order, harmony, and incorruptibility—utopia.

Leibniz's calculators were constructed from the components the seventeenth century had to offer: wheels, axles, cranks, and rollers. Modeled on the ten fingers of the human hand, the gears in his calculator had ten teeth. Sums could be added by turning the wheel. One such machine was acquired by Czar Peter the Great of Russia, who was keenly interested in the transfer of technology from Western Europe.

The next great impetus to mechanized calculating came with the wave of industrialization in the nineteenth century. The need for rationalization and cost reduction in the textile industry, as well as an increase in the number of woven patterns, had led to the introduction of the Jacquard loom. On these looms perforated cards with individual instructions for the pattern allowed the warp threads to be lifted in the proper sequence to weave complex patterns. Charles Babbage, a British engineer and mathematician, transferred this control principle to mathematics. His "difference engine," presented in 1822, was intended to predict the tides, currents, and perhaps even the weather. The machine was complicated to use, but not totally impossible. Its predictions were inaccurate only because so many influences had to be taken into account. Scientists continued to believe that powerful, automatic calculating machines could be made.

Babbage also worked in insurance, an important new business in England, catering to the growing world trade. Statistics seemed the only real means of calculating risks and opportunities. Statistical estimates, however, require large amounts of data to keep from ascribing too much importance to occasional, isolated deviations from the norm. The automatic manipulation of large quantities of figures would enable businesses to effectively manage international economic dealings. Babbage spent more than thirty years trying to convert this idea into a practical machine. He wanted his machine to use steam power, a novel form of energy in his age, but there were too many difficulties. Nevertheless, Babbage's plans for an "analytical machine" were based on useful concepts. For example, he differentiated between a storage unit for his data and the control device needed for calculating. Babbage already realized that a machine of such a size had to be constructed using mechanical parts and subsystems with a great many cogwheels, axles, and levers, one system following the next in long repetition. A small error in the calculations would rapidly become a large error within these successive mechanical systems unless each cogwheel and component were made with great precision. Standardization of parts seemed necessary, yet Babbage sadly realized that his standards could not be met by the mechanical engineers of his day. The accumulation of errors in the propagation of calculations through an "analog computer" of this sort was even then seen as a major problem. Scientists had to wait another hundred fifty years for the symmetrical atomic structure of semiconductor crystals in order to make reliable computing machines.

When electricity came along, it offered great improvements over mechanical energy sources such as steam, gas, and hydraulic power. Because the flow of electrons through a conductor does not result in any lasting alterations and leaves no waste, electricity seemed ideally suited for computers. But what kind of switches could be used? How could different states symbolize numbers and instructions in a recognizable manner? Only a mechanical instrument seemed able to solve this problem. So the first computers used an indicator, yet another mechanical device. For decades the combination of electricity and mechanical processes was the only practical way to make large computers work.

A German civil engineer, Konrad Zuse, followed this route. Zuse, who was a skilled engineering designer, draftsman, and artist, was irritated by the drudgery of arithmetical processes. Technical drawing was already governed by such strict procedures that everything it required could be automated. Even more wearying were all the calculations essential to any well-planned construction project. Zuse spent hours speculating and planning a computer that would spare people needless effort. In 1936 he shocked his parents by quitting his promising position as a statistician to set up a home workshop. It was to be the birthplace of the first real digital computer.

Zuse used telephone relays, controllable mechanical switches, as structural elements to build his Z3, the world's first truly functional computer, completed in 1941. It was a program-controlled computer having a storage unit interconnected with 1,400 relays. At first the German war effort helped promote his undertaking and provided some political support; a machine that could easily compute the aerodynamic resistance of a fighter plane, which required large numbers of calculations, would be of great use to the military. But the war caught up with Konrad Zuse; most of his instruments were destroyed in the Berlin air raids.

At the end of the war Zuse was taken prisoner and sent to London to be interrogated by an official of the British Tabulating Machine Company, a competitor of Zuse's computer company, which he had started in 1941. The official did not speak German, and Zuse spoke only a little English. The conversation did not shed any light on Zuse's achievements, and for many years his contributions were simply overlooked. Zuse's company revived after World War II, then had severe difficulties in the sixties; in 1967 it was absorbed by Siemens, the giant electronics firm. Even today, many works on the history of modern technology make no note of Zuse. Not until 1965 did the American Federation of Information Processing Societies award him a medal.

In the English-speaking countries the war meant great advances in the development of the computer. In 1936 Alan M. Turing, a British mathematician working at Princeton, had written a fundamentally important paper whose title contained an interesting mixture of languages: *On Computable Numbers, With an Application to the Entscheidungsproblem*. Turing's work discussed the construction of a program-controllable computer. Turing, an eccentric fellow, often wore a gas mask when he rode his bicycle, to prevent hay fever attacks. In addition to being the top man working on the principle of the electronic computer, Turing was a long-distance runner and the author of poetical fairytales. Electronic computers today still operate according to his simple and convincing principle. A movable paper strip, containing only zeros and ones to convey the operating instructions and the data to be manipulated, showed that a computing machine can perform all necessary operations.

During the war the British military became interested in the calculating machine. They wanted to outdo the German army's Enigma, a sophisticated encryption machine that was able to mechanically scramble the high command's orders to the troops. A receiver built according to the same mechanical principle then was used to decode the messages. Britain's answer was an electronic computer named the Colossus, which used countless electron tubes to numerically imitate the Enigma.

Turing's basic concept paid off, and the German military code was revealed, a mathematical triumph of major strategic importance for the British.

Artillery and ballistics provided another use for mathematical calculations. France's golden age of mathematics owed much to Napoleon's promotion of the science for artillery research. In the United States, even before World War II, Herman Zornig was appointed to head a research center to determine the trajectories of grenades. This important laboratory was later led and advised by top researchers, including some Nobel Prize winners. A typical ballistics curve computation, a not too difficult task, requires about a thousand arduous multiplications. A firing table, listing all the possibilities, requires the computation of several thousand individual trajectories. Even with a mechanical calculator, it took twelve hours to determine just one trajectory, and the mechanical gadgets were often cranky.

The U.S. military found that it still took too long to calculate the trajectories required for quick air defense. To come up with the answers, they brought together artillery science and mathematics. One of the most dependable calculating instruments of the time could work all night long without supervision. This machine operated at Bell Telephone Laboratories and worked with standard telephone equipment: mechanical relays. In 1944 Model III was finished. It had 9,000 switches, took up more than a hundred square yards of space and weighed ten tons. It allowed two seven-digit numbers to be added in a third of a second and divided in two seconds. It was now possible to calculate a trajectory in less than forty-five minutes.

But as the war progressed, it became increasingly clear that electromechanics would remain a slow process. The University of Pennsylvania was commissioned by the ballistics experts to work on the problem, but only after the war did they come up with a truly functional large-scale computer, named ENIAC. This computer dinosaur, used for atom bomb research, required no fewer than 18,000 electron tubes and 30 tons of equipment housed in a large room. The machine consumed 150 kilowatts and generated a huge amount of heat that was difficult to vent. Yet it could compute more than a thousand times faster than any electromechanical device. ENIAC computed just as young children do, counting on all ten fingers. Zuse, however, had already simplified this system, as did many later researchers. In due time scientists learned to count using just two fingers, and the digital age began.

The Business of Binaries

The system of counting with ten digits—0 to 9—came about only because nature gave us ten fingers. Other bases are also possible, such as counting by dozens or fifteens.

Long before the first computers appeared, mathematicians had developed the method of counting with just ones and zeros, but this base-two, or binary system appeared too artificial and theoretical to be generally useful, compared with the decimal system. The binary system permits any number to be represented by a row of zeros and ones. Every one means that a power of two—2, 4, 8, 16 (and so on), depending on position—is to be included in the sum. A zero means that no number is in that position, just as the 1 in 100 means that we have a single hundred, no tens, and no ones.

The binary system differentiates between two cases: yes, or one, and no, or zero. This simple choice, which can be used to designate letters and symbols as well as numbers, allows a wealth of possibilities. The alternatives of yes and no can be easily incorporated into an electronic circuit with an on (one) and an off (zero) switch. When using the old ENIAC, scientists had to divide a current into ten steps for the individual digits. A simple on/off decision is much more clear-cut than an electrical current representing a 7, say, or an 8. The current representing a number could change as it traveled, so there was a risk of inaccuracy. Using the binary system, a series of on and off electrical switches can be used to represent any number. This requires long strings and many components, which reduced speed, yet the principle proved to be much more logical and precise than the base-ten system.

The simple operations of adding and comparing were easy with the system devised by the English mathematician and logician George Boole. Logic that could simulate "and/or" states was easy to create; a simple two-input switching device would forward an output signal only when both inputs were on. The output signal would appear as the message that input 1 and input 2 were on. All the rules were there. Toothed gears and finely adjustable electron tubes did not need this kind of simplification, but these complex components had to be extremely accurate to keep errors from accumulating in the course of the calculations.

The transistor seemed perfectly suited to this binary system. Even in Ferdinand Braun's time, the semiconductor crystal had appeared particularly suited to differentiating between "on" and "off." It was the crystal's nonlinear behavior that had so surprised Braun. Linear behavior means that a little more input will generate more output, in exactly the same proportion, as given by a straight-line relation. Nonlinear behavior arises when output is either/or, on versus off. Below a certain threshold, a little more input will not change a zero output. Above the threshold, a little more input always results in an output of one. This seemingly naive concept of zero output below and one output above a threshold has a distinct advantage. This yes-no algebra can be done with simple on-off switches, and, even more important, errors are suppressed efficiently. Braun's crystals permitted current to flow in one direction; in the other, flow was blocked. Even the simple setup of a crystal and a tip showed obvious differentiation between on and off.

Scientists today distinguish between two techniques of representing quantities and values. The "dual" method, using semiconductor microelectronics, led to digital technology. The older process, called the analog technique, reigned supreme until the 1950s. When a loudspeaker reproduces a certain volume, the strength of the electrical current is analogous to the intensity of the sound. The density, or blackening value, of the little crystallites in the photosensitive layer of a photograph are analogous to the amount of light shining on the layer. The mercury in a thermometer rises to a level analogous to the temperature. A clock depicts an angle of the hands that is analogous to the amount of time that has passed. The digital system, by contrast, is a process of counting and encoding numerical values. The primitive code recognizes only two symbols and therefore produces a long chain of figures. Yet such chains are easy to transfer and process. More important, they are readily and securely stored in systems that can differentiate only between "yes" and "no," on and off.

The transport of information via an analog channel is jeopardized by unavoidable chance fluctuations and foreign influences. The communications engineer calls this "noise." Noisy transmission channels sometimes distort analog signals beyond recognition. Digital systems have procedures for at least recognizing, and sometimes rectifying, this problem. Given this advantage, binary representation is being used more and more to store information. Compact discs, which are based on binary representation, are the precursors of increasingly digitalized electronics for sound transmission. A digital compact disc covered with greasy fingerprints or handled carelessly will still provide amazingly good fidelity; a normal long-playing record demands kid-gloves care.

Any piece of information in the form of an analog signal can be converted into a digital chain of zeros and ones. The changes of pitch and intensity of music, for example, are scanned at regular intervals and the values converted into a series of digits. The more frequently this scanning takes place, the higher the quality of sound reproduction. Speed of translation is decisive in applying digital technology. The smaller and more maneuverable a zero-one switch is, the more varied its uses. The changeover to digital techniques had to wait until the crystal could provide many little rows of quickly reacting switches lined up one after the other. Digital technology and modern microelectronics thus grew up together.

The path to today's digital technology was long and hard. After the war many Americans were clamoring for larger computers. The need for accurate weather forecasts and for the processing of large amounts of data from economic surveys and censuses gave a push to development. Thomas J. Watson, for one, whose company, International Business Machines, manufactured simple office machines, foresaw an important role for electronic computers in the workplace. Watson commanded his workers to "Think!"—think of a way to corner the underdeveloped market for automated office machines. IBM specialized in devices for reading and storing information from little square perforations in cards. These computer cards, which had evolved from the programmed Jacquard loom cards, were ready to be replaced by electronics. "Just imagine," Watson is reported to have told his employees around 1930, "only two percent of office work is done by machines. What a great field for work and opportunity." Microelectronics was to give his firm its big break, enabling it to surpass all its competitors despite their occasional considerable leads. Today IBM controls about 70 percent of the computer market and is a leader in technology. A German industrialist once admitted to me the secret of IBM's success: research right down to the grassroots; solid, secure technology; and an inimitable network of technical and business competence in customer service.

Research, research, and more research was imperative to change the computer's system from tubes and relays to germanium transistors. It was not easy, and the first solutions were clumsy and awkward. The long chain of germanium elements lined up one after the other resulted in lengthy current paths. The whole computer seemed to consist solely of connection wires, with transistors almost unnoticeable. And it seemed almost impossible to use semiconductor crystals for the memory part of the computer. There are two main parts to an electronic computer: logic and memory. The logic section consists of combinations of elements that permit the numbers to be computed, compared, and manipulated. In the storage or memory unit, data must be filed, and the computed results stored as data. The operating instructions also must be stored in such a way that they can be recalled quickly. The semiconductor crystal managed to conquer logic switches; memory, though, proved resistant. To store data, a computer requires a tight configuration of many small cells. These cells store an electric quantity in an unambiguous and permanent form, so that each request for information is answered with a definite yes or no. The specialist calls this requirement bistability. Only two states—and nothing in between—are stable; the machine switches back and forth between them.

But germanium had great difficulties fulfilling this prerequisite. Each cell needed several transistors and resistors cleverly interconnected, as well as connecting lines between cells. The bulky setup occupied more space than could be allowed for in a storage unit, and it slowed down operations. A single germanium crystal was unable to store signals unambiguously and for long periods. Electrical charges might have been used to symbolize zero (no charge) or one (charge), but great effort is needed to fix and preserve the charges. The proper instrument for storing electricity had not yet been invented, so the best solution seemed to lie in connecting several transistors. Together, they were an acceptable substitute for a mechanical switch that could differentiate between "on" and "off."

The semiconductor crystal simply was not up to the task of storing information. Electricity was defeated by its sister phenomenon, magnetism. A magnet is bistable; the polarity of a magnetic material can be changed back and forth. At a single spot on a magnet there may appear first a magnetic north pole, then a magnetic south pole. And small magnets can be made. Minute, flat rings of a magnetizable material can be threaded on a wire, just as beads are strung on a necklace. Electrical current can be used to reverse the polarity in the magnets and thus write in information. And the information inscribed through the polarity of the magnet can also be read out via wire feeds and reconverted into electrical signals, zeros and ones. These magnetic storages looked like carpets. On a fine wire network stretched between stable frames, the tiny magnetic beads containing the digital information dangled at each intersection of the wires. This technical solution was sound, and was used by IBM and other companies until the sixties, but engineers could not generate much enthusiasm for it. Inordinate amounts of electrical energy were required to reverse the magnetic field, and manufacturing these wire mazes was a terrible job for the assembly-line workers, even when machines aided in stringing the beads. The whole setup was too reminiscent of traditional technology: workers were merely assembling prefabricated components. Magnetic storage units suffered from the same disadvantages as vacuum tubes composed of many individual parts.

No one had figured out how the crystal could store information, as well as act only as a switch and amplifier. New theories, perhaps even new materials, were needed to create a functional, multipurpose semiconductor.

6
Settling in Silicon Valley

I couldn't say I hadn't been warned. Most of my coworkers and friends thought I was crazy to leave my post as a scientific assistant at a German university and head for California. Even worse, I was not taking a job at a respected university like Stanford or Berkeley; I had decided to work for a very small industrial company, where my pension claims were still in question. Why leave Göttingen for this madcap adventure? The only point in favor of the plan was that my new boss was Nobel Prize laureate Bill Shockley. But shortly before my departure, Robert Pohl, now an old man, and John Bardeen, who was visiting, came to visit me in the lab. Bardeen looked at me for a long time with his melancholy eyes and finally said, "Well, we'll see how you like working with Shockley."

Thus warned, I made my way in the fall of 1959 to 391 South San Antonio Road, Mountain View, California. Before going in, I studied the building in disbelief: this old apricot barn was where a Nobel Prize winner had his laboratory? There was not even a decent sidewalk and no real front entrance. But the work being done here was to turn the fruit orchards of Mountain View and Palo Alto into a new industrial landscape: Silicon Valley.

Technology Moves West

After receiving the Nobel Prize in 1956, Bill Shockley reportedly said that he had seen his name often enough in *Physical Review*; he now wanted to see it in the headlines of the *Wall Street Journal*. The theoretical physicist turned down all offers to teach at universities, remained totally uninterested in a secure, prestigious position at Bell Laboratories, and instead borrowed enough capital to finance his own firm. He took time deciding on the firm's location. The East Coast—New York or Boston—was out of the question because it was too conservative and not attractive enough to lure good scientists. Florida? He needed an academic climate and a cultural atmosphere. He considered the California Institute of Technology, but he did not care for the smog around Pasadena. The peninsula south of San Francisco, which included the little university town of Palo Alto, seemed the best location.

In addition to Stanford University, Palo Alto already had a tradition of technology. Much earlier Lee De Forest had moved to that thinly populated region of fruit groves to develop the electron tube, and the Varian brothers, also working with tubes, had more recently built up an electronics firm of considerable stature. And in 1938, William Hewlett and David Packard, working in Packard's garage in Palo Alto, had developed an electronic instrument that they proudly named the Type 200 A Oscillator. This impressively named device was the start of a large company that manufactures precise measuring instruments. Frederick Terman, dean of engineering at Stanford, had encouraged the two young men to take on the entrepreneurial challenge.

Following the pioneer tradition of the West, Stanford University in 1951 decided to attract new scientific industries to the area. Part of the university's endowment is a vast tract of land, which may not be sold but may be rented to tenants dedicated to scientific endeavors. The firms that move to the Stanford Industrial Park must create an academic, but not cocoonlike, atmosphere. They are also required to take part in academic projects for the university. In exchange for helping to finance learning endeavors, the tenants are given access to Stanford's research facilities as well as to potential employees. The tenant-company buildings are mostly modern, and some are quite daring and avant-garde in their architecture. Well-tended gardens abound, and the contracts contain environmental protection guarantees. This industrial park concept was begun at a time when few Americans were thinking about ecological awareness.

When Shockley moved to the area in 1955, first into an inexpensive barn rather than a fashionable lab in the industrial park, his interviews with young scientists were rigorous; he required psychological, as well as academic, profiles of all candidates. Arnold Beckman, an old friend who owned a successful company that manufactured instruments, financed the founding of Shockley Semiconductors as an affiliate of his own company. All the ingredients of this new enterprise in crystal electronics spelled success: the well-known researcher, the location, secure backing, and ready markets for new products. But the development was like that of a Greek tragedy, in which hubris and belief in the oracles cause the protagonist's downfall. Shockley helped create a whole new industrial world but failed at his own venture.

While still at Bell Labs, Shockley had invented a diode, a circuit element with only two connecting leads that demonstrated bistability—that all-important dual capability for yes and no. Disappointed that the transistor had not been named after him, Shockley hoped that this new device, his company's first product, would immortalize him. This "Shockley Diode" was a beautiful, subtle element, with three *p-n* junctions inside the crystal. Its method of operation was highly sophisticated and difficult to comprehend. As a young scientist at Bell, Shockley had been captivated by the idea of inventing a simple switch for the telephone system, and he thought his invention could be used as a telephone component. When his old friends at Bell seemed skeptical, he undertook the project alone.

Shockley found that the diode had to be made of silicon; germanium had proven ineffective. Scientists the world over, especially on the American East Coast and in Europe, had hedged their bets on germanium as the basic crystal material; it was convenient to use because it could be melted and alloyed at fairly low temperatures, almost like a metal. Brittle and chemically aggressive, silicon was not nearly as pleasant to work with, and its melting point is above 2,500 degrees Fahrenheit. To obtain a suitable crystal, silicon must be heated to extremely high temperatures in a quartz crucible. Why go to such pains if germanium worked well enough? But germanium had already shown its limitations. Its close relationship to metals created problems. Solid-state electronics requires a semiconductor, not a metal, because it is not possible to externally influence the number and distribution of electrons in metal. If a powerful current makes the germanium too warm, it releases so many electrons from its crystal bonds that all the painstakingly prepared functions are drowned in a sea of electrons. In Europe the radios of cars parked in the hot sun sometimes stopped receiving broadcasts because heat incapacitated the germanium transistors.

Because of such problems, the big electronics firms replaced germanium with silicon in devices using strong current. Unlike germanium, silicon resists heat and holds on more tightly to its electrons. It remains a semiconducting crystal for a much longer period and retains its functions as determined by the donor and acceptor atoms. It is much better suited to switching and high levels of electric current.

Germany was the first country to apply silicon technology to transistors. In a converted castle in a village in Franconia, Walter Schottky's research and production team produced large silicon rectifiers and switches for the Siemens Company. The "Siemens process" for manufacturing ultrapure silicon was a groundbreaker, and today it is licensed around the world. Under the auspices of the Wacker-Chemitronic Company in Burghausen, West Germany, still leads in the production of the chemical base material.

Silicon (from the Latin *silex*, meaning flint) is one of the most common elements in the earth's crust; it is all around us. We do not find silicon in its pure form in nature because it forms a stable compound with oxygen; sand is nothing more than the compound of one silicon atom with two oxygen atoms. Entire mountains consist of minerals called silicates, which contain oxygen and silicon. To prepare pure silicon, the silicon-oxygen compound first has to be violently and energetically separated into its components. The resulting fine crystalline powder has to be purified and formed into a perfect, single crystal. The process is long, difficult, and expensive. Nevertheless, silicon today has cleared the field of all its competitor crystals.

As Shockley and his young troop began to work with silicon, they kept in touch with "Ma Bell" in New Jersey, exchanging knowledge and information. Bell allowed Shockley to use its patented processes. Perhaps, they thought, the mighty Bell Telephone empire would buy up the Shockley diodes and install them in hundreds of thousands of switchboards. But the business was going very badly. Especially disappointing were the unreliability of the diodes and the low yield. Yield, the percentage of usable, salable, devices from the total number of those fabricated, is the most serious problem in semiconductor electronics. An industry such as automobile manufacturing usually achieves a yield close to 100 percent. But the batting average in semiconductors was often terribly low; yields below 20 percent, for example, were not at all uncommon. No other technology ever had to accept such frugal yields. When a new device is first processed, only a few specimens are up to standard.

Silicon technology proved to be incomparably more difficult than that of germanium. In addition, the physical mechanism of bistable switching was more complicated than that of a transistor. Shockley worked hard testing new alternatives to make his new device switch correctly. Mechanical switches and relays were easier to design and assemble.

A new, more precise technique, called dopant diffusion, replaced the alloying process for doping crystals with donor or acceptor atoms. In this method a piece of silicon crystal is pushed into a quartz tube inside a furnace heated almost to the melting point of silicon. At that temperature the crystal lattice relaxes and allows foreign substances to enter. One of two kinds of foreign element, an electron donor, such as phosphorus, or an electron acceptor, such as boron, is then vaporized and allowed to enter the quartz tube. Exact temperatures, calculated to a fraction of a degree, and precise heating times in the furnace are required to create *p-n* junctions and establish their depth. The hot diffusion furnace has become the heart of every semiconductor factory, providing the crystal with form and function.

In spite of the new technique, Shockley's diode remained highly problematic. Most of the finished ones had to be thrown away; only a few had the right electrical properties after the diffusions. Fortunately, manufacturing with crystals requires only small amounts of material; they could afford to throw away a lot. The next step was to improve technology, increase yield, and reduce costs and price. But research was becoming increasingly expensive for Shockley's team in the apricot barn in Mountain View. The market was not growing quickly enough; despite their hard work, success evaded them.

Mutiny in the Laboratory

In the meantime, in the late fifties the team did learn some better methods for working with silicon. At day's end, watching the beautiful red sunset, they would discuss how to ward off the impending failure of the new business. What new marketable product could they create? Why not ordinary transistors? Although that idea seemed silly, almost old-fashioned, the Shockley team decided to conduct some secret experiments with the furnaces. They knew that there appeared to be a market for good transistors. Still shocked by the Soviet Union's Sputnik triumph, the United States was eager to regain the lead in space exploration. NASA and the military were interested in microelectronics for use in rockets and satellites and were willing to pay any price for reliability and resistance to temperature fluctuations. But Shockley was hurt when his team proposed dropping the diode project, and he flatly refused to give it up. Seeing no compromise possible, the young team mutinied.

Robert Noyce, a trusted and good-natured team member, became the rebels' spokesman. The son of a midwestern preacher, Noyce had been a hobby craftsman and an amateur scientist as a boy. While still a physics student, he had worked with the first transistors, and he was fascinated by the theory of the new semiconductors. After receiving his doctorate from the Massachusetts Institute of Technology, Noyce had dedicated himself to germanium technology until Shockley lured him to California. According to Silicon Valley rumors at the time, Noyce's psychological test revealed that he was a brilliant physicist but not a good manager. Ironically Noyce today is known as a very successful scientist-manager. The silicon firms he has founded have earned him at least sixty million dollars.

One day a representative of the mutinous group from Mountain View gave Shockley's financial backer, Arnold Beckman, a report on the possibility of using silicon crystal technology in transistors. The representative was sure that improved transistors could put the failing firm in the black. Beckman listened attentively, knowing that a revolution was in the making. He believed that entering the world of transistors was risky but not hopeless. But Beckman decided that this decision would amount to a stab in the back to Shockley, his friend and partner, and probably would be a breach of contract, so he turned down the proposal. It was now impossible for these men to continue working with Shockley, so eight of them, including Noyce, left the lab. Feeling forsaken and deceived, Shockley called the men "eight traitors." Depression crippled him for months.

The young turncoats had to decide what to do. Should they go back to peaceful, secure jobs on the East Coast? Should they look for work in the labs of established concerns that used germanium, and give up on brittle silicon? Noyce, just twenty-nine, an age when most European physicists are starting to look for a thesis topic, refused to throw in the towel. Instead he looked for capital to start a new firm to be owned and operated by the team. He found Hayden Stone, an East Coast investment broker, who had been working with Fairchild Camera and Instruments. Fairchild was interested in making aircraft parts and wanted to use the new semiconductor technology. The mutineers presented Fairchild's president, John Carter, with a risky but well-thought-out plan. Carter agreed to finance the group as an independent company, called Fairchild Semiconductor Corporation, but reserved the right to take it over after two years.

When Fairchild began in Palo Alto in mid-September of 1957, the average age of the team was less than thirty. Bob Noyce and Jay Last took charge of photolithography. Vic Grinich and Murray Siegel, a new member who had originally intended to work for Shockley, headed the applications lab, which planned and tested the devices. Sheldon Roberts had the task of growing sufficiently pure silicon crystals. Gordon Moore and Jean Hoerni, a Swiss immigrant, brought along the techniques of donor and acceptor diffusion, which they had learned in Shockley's barn. Eugene Kleiner managed the business and organizational end.

This secession marked the first of many new companies that branched off from existing semiconductor firms. In 1968 alone, thirteen new spinoff companies arose from Fairchild. That same year Charlie Sporck, the hard-boiled factory manager, left and founded National Semiconductors, a company that reached billions in sales, and Noyce took a few friends from the firm and established Intel. Hanging on the walls of many Silicon Valley offices are company "family trees," which each year sprout a new growth of firms that can trace their ancestry back to Fairchild Semiconductor. That company, affectionately called by many names, including The Ranch, The Flying F, and Mother Fairchild, was a reservoir of talent for many new firms until it was finally taken over in late 1987 by one of its offspring, National Semiconductor, which paid the pitifully low sum of 122 million dollars for it. And Fairchild's roots, of course, go back to Shockley's old apricot barn.

In the late fifties conditions for launching a company were more favorable than when Shockley began. After Russia's launching of Sputnik, politicians and military leaders raced to improve the teaching of science in the schools. The military discovered outer space as a possible battlefield. They needed computers to calculate rocket trajectories and orbital ellipses. They wanted electronic equipment for establishing communication between the earth and satellites. The equipment had to be lightweight and sturdy, able to withstand rapid accelerations, and impervious to heat and radiation. Germanium could not meet these requirements; silicon, that brittle, resistant crystal, fit the bill.

In this hectic, frantic phase, Fairchild's first product, the silicon transistor, proved to be a market success. What the turncoats in Shockley's barn had envisioned became a reality. Though problematic, silicon technology could be effectively precalculated and planned. First a photo etching method adopted from printing technology was used for dimensioning, that is, defining the structure of the transistor. In a darkroom the silicon wafer, ½ millimeter thick, which was sawed off the crystal, was coated with a photosensitive resist. Using a fine mask (a black and white glass plate), the pattern of the transistor was transferred onto the resist as a photographic image. Wherever light shone on the resist, it became insoluble. The unexposed areas remained soluble and were removed with a liquid. The next step called for hydrofluoric acid, a very powerful and chemically aggressive acid that will attack even silica, the oxide of silicon. Hydrofluoric acid penetrated the oxide layer on the silicon. After etching, those areas not covered by resist were bare, unprotected silicon crystal. In the furnace, electron-donating phosphorus atoms entered the crystal at the unprotected areas.

Traditional photo etching techniques, combined with suitable oxide resists, permitted scientists to choose the exact spots where they wanted to change the crystal. The old methods of using a gun to spray black wax through a mask now seemed ridiculous. Even more important was the diffusion method of adding foreign atoms by which one could control the depth of penetration within thousandths of a millimeter. The new system was ten times more precise than the alloy method of germanium technology. From that time on, every idea and every skill focused on better regulating dimensions on the surface and in the interior of the silicon crystal. Microelectronics had become truly "micro."

The first silicon transistor was given the romantic Spanish name "mesa" because etching caused each one to look like a tabletop mountain. The robust mesas could withstand strong currents. They were nimble; a current inside one could oscillate up to one billion times a second. No wonder they were high-priced—up to one hundred dollars apiece. Only the military and NASA could afford such high prices, and military contracts helped Silicon Valley get its big break. In 1959 and 1960 almost half of the total dollar value of transistor production went for military and space purposes, even though the actual number of transistor pieces bought for such use amounted to only about 25 percent of total production. The government demanded expensive, powerful silicon. The death knell had rung for germanium.

While noting the large U.S. government expenditures, we should not overestimate the role of military research in the development of microelectronics. Europeans usually overstate the importance and size of the U.S. military market and use it to explain the American lead in this sector. Government support was important only during the early stages, especially for the Valley's financially weak novices, who needed cash immediately. Many of them took advantage of government financing of research and development. The established, powerful firms, though, preferred to make it on their own. By the mid-sixties, silicon product purchases by the military and the space industry amounted to less than 25 percent of total sales and just 10 percent of the units sold.

The strict technological requirements of the government customers lent wings to the new industry. Housed in its little brass pot, the transistor had to withstand heat and cold without altering its characteristics. It had to remain unaffected by submergence for hours in salt water, as well as by the rattling and shaking of a rocket being launched. Civilian radios or hearing aids did not have to meet such stringent requirements, nor could they afford the high price of the silicon transistor. The military's insistence on strenuous tests for high quality forced transistors to become reliable. Only this "milspec," or military specification guarantee, enabled the transistor to later conquer civilian markets.

While the eight runaways were making their first profits at Fairchild, Shockley pulled himself together. His company, renamed Shockley Transistor, sought new scientists. To me, a young German from Göttingen, who had never even heard about transistors or semiconductors at the university, it sounded like an interesting opportunity. Only fifteen minutes after I landed at the San Francisco Airport, the colleague who was driving me along Bayshore Freeway to Mountain View told me that Shockley Transistor could go bankrupt the following week. But not to worry, he said. I could always go to Fairchild down the road—they always needed semiconductor people. Or if not that, well, somebody would surely buy our little company and keep things moving along.

And that is just what happened. Scouts from all kinds of East Coast firms snooped around in our lab in the barn. A company that had earned a lot of money on the cheapest germanium diodes, and was terrified of the threat of silicon, finally purchased Shockley and his boys, the patents and the furnaces, silicon crystal technology—even the old barn. They constructed a new building at a beautiful location in Stanford Industrial Park. Good work was done, but the company was not as successful as the competition. It was again sold as a package deal, and during the crisis around 1963, the firm finally gave up. When the team split up, I went to the East Coast, to the mecca of Bell Laboratories.

Working with Shockley had been both demanding and rewarding. His extremely quick grasp of scientific problems made him a very impatient boss. A suggestion to test one of his new theories would be followed a day later with questions about what the experiments looked like. His strategy was to get to the heart of the matter very quickly. Fuzzy thinking was reprimanded; you were never supposed to say that a transistor had "a short circuit" but to specify the exact value of the resistance of this connection. When I first arrived, I was asked to start on a silicon solar cell project, and out of that came a joint paper on a truly basic theory of solar cells. Many other projects followed, including one on the influence of dopant atoms with radii smaller than those of the host silicon atoms, which therefore dislocate the silicon lattice plane. Once I established a relationship of trust with Shockley, it was a strenuous joy to work with him; his critical guidance conveyed the true spirit of scientific work.

War and Peace

Silicon Valley continued to spread out through the peninsula south of San Francisco. Cherry trees, fruit groves, and marshes gave way to laboratories and factories. Thousands of people were drawn by the magical success of the new industry. Sensational stories dramatized the profitability of silicon crystals. Many people arrived scheming to get their bearings for a few months, then find like-minded colleagues and set up their own companies. They brought knowledge and customers from previous companies and set about competing with their former employers. A totally new climate arose—a mixture of optimism, competition, mistrust, and frantic activity. The fertile fruit valley turned into a blossoming economic region, accentuated by quick profits and shaky loyalties.

As the demand for academic research and theory shrank, the university played less and less of a role in the industry's scientific training and education. The entrepreneurs wanted engineers who could quickly convert the physics and chemistry of semiconductor crystals into improved manufacturing techniques and products. By attending the engineering conferences at which Bell and IBM scientists lectured freely on their basic research, anyone could learn about the latest developments and then apply them. As the production of silicon transistors rose, craftsmanship improved. The production lines, with their giant furnaces, became more important than classrooms and laboratories. The Wagon Wheel Bar, located strategically close to Fairchild and its spin-offs, was the place to make contacts, trade information, and start rumors.

Every spectacular success story increased the appeal of the Valley. Even failures won admiration and respect. Company loyalty and security were rare in the new industry. Bosses had to be cautious in dealing with subordinates, for they might be heading a competitor firm next month; you might even have to ask *them* for a job some day. Traditional hierarchies gave way to unusually young teams united in their rejection of the East Coast establishment.

The Valley scored a big victory when Fairchild had to hire a new head from outside. The new boss would accept the position only if the headquarters of the parent company on Long Island moved to Silicon Valley. This move confirmed the West as the center of new technology. The East Coast companies were more stable and financially stronger, with good research labs and the power to control politics and markets. The West Coast could maintain its lead only if its more flexible structure led to continual technological progress.

The financing of new firms was a major problem. In the beginning, investments were small and risks could be calculated. Each year of technological progress, though, raised the entrance fees and increased the risks. On the other hand, every success story reassured those who wanted to invest in risky electronics ventures. A capital market developed far from the official Wall Street and Chicago stock exchanges. Small-scale, skillful brokers advised investors when to leave and reenter a field or market. Eventually these financial consultants made headlines; one of them, Arthur Rock, became the head of the Silicon Valley investors. Paper profits amounting to millions were possible. Some companies disintegrated immediately, but others lasted. Headlines on the cover of *Time* magazine announced: "Cashing In Big—The Men Who Make the Killings." The *Palo Alto Times*, an unusual paper filled with technical and financial news, found it hard to keep up with all the new enterprises. Yet it was a small world, and most news traveled quickly by word of mouth.

In the early stages the new firms did not have an easy time. Salaries were low, even for the founders and the top people. The main incentive to employees was not income, fame, or even pension plans but the company stock. The prime mover of the Valley was not a high salary, but the chance—and risk—of becoming a millionaire. Thus many investors bought stock in unknown technology firms, hoping to make a killing, which, under certain conditions, could be practically tax-free. Some of the small businesses grew quickly and returned threefold to tenfold returns on investment in their early years.

Plane and Simple

The high spirits felt during the early stages, the thrill of taking a risk, the pride, and the drive to work, work, work—all of these spurred on technology. The silicon crystal learned to do things that even the experts would have thought impossible. The problem with the mesa transistor etched from a crystal wafer was its hypersensitivity. Because its *p-n* junctions were not protected from the surrounding air, the slightest amount of water vapor, the tiniest bit of dirt in the atmosphere could change the flow of electrons and distort switching and amplification. Even though the transistor was pressed into its little brass pot with pure nitrogen gas, its surroundings could still affect it. It was the same problem that had plagued Braun: the difficulty of controlling what happens on the surface.

The eight founders of Fairchild knew that the fate of the new company hinged on mastering this problem. The *p-n* junction had to be sealed and protected inside the crystal. What finally saved the day was that an incredibly stable oxide of silicon can be wrapped around the crystal to protect it. Oxygen fed into the hot furnace makes a protective covering on the silicon wafer. Germanium, by contrast, cannot easily be given a stable oxidized layer; it is necessary to introduce foreign atoms, which are responsible for all the disturbances in the first place. Silicon's oxide permits making the transistor in a flat shape, a planar transistor setup that does not leave any sensitive parts unprotected. "Planar technology" using passivated silicon surfaces resulted in great improvements for Fairchild and the industry as a whole. Yields rose, prices dropped, and "silicon performance at germanium prices" was a promise that could be kept.

Crystal processing changed. Before, technicians had to manipulate the entire volume of the crystal, penetrating the interior by etching, alloying, and separating. Now only the surface of the silicon was processed, and this surface remained protected. The foreign atoms that create the *p*- and *n*-zones diffused into the crystal through minute openings etched into the oxide. Metal vapors formed the paths for current within the crystal, while leaving the interior unexposed. The protective silicon oxide layer was also a separator or isolator, allowing technicians to determine the exact points where electric currents should enter. The nonconducting oxide made sure that no unwanted circuits occurred elsewhere. The new planar technique affected only the thin layer of silicon just below the protective oxide. The rest of the crystal was now merely the handle for the surface area. The current simply journeyed unimpeded down to the substrate and then exited.

Compared to germanium, silicon required greater caution and more sterile conditions, but these preparations paid off. This brittle element, with its high melting point and perfect protective oxide, gradually became *the* semiconducting crystal. Silicon wafers were made thinner and thinner; only their brittleness required that they be at least a fraction of a millimeter thick. The new technique led to ever faster, more reliable, and smaller transistors. The single crystals drawn out of the hot crucible became larger and more symmetrical. In the early days of the technique, cylindrical single crystals measured one and a half inches in diameter; by the mid-1980s, they were more than four times that size. The material in which the Silicon Valley alchemists buried their electronic functions came closer to perfection.

Another new technique in crystal preparation involved applying a vapor containing silicon atoms to the crystal. The silicon atoms precipitated onto the surface and assumed their proper locations in the crystal structure. If no contaminants disturbed the process, a new piece of crystal with the required electrical properties grew slowly and symmetrically. This production method, called epitaxy, shortened production time and increased flexibility.

In the early 1960s scientists wondered whether the techniques of nuclear physics could be used to inject the donor and acceptor atoms at high speed into the crystal. The diffusion process took too long and was jeopardized by contaminants in the white-hot furnace. Nuclear physicists scaled down their accelerators and used them to bombard the crystal with atoms. The massive broadsides distorted the crystal, but a short heat treatment restored the symmetry of its lattice. This was another step forward for crystal technology, but it raised the cost of processing.

As Silicon Valley grew, property prices rose and population density increased. Air pollution became a problem. As we drove to lunch one day in his green convertible sports car, Shockley rightly predicted that the smog of southern California would also afflict the North as population density changed the landscape of the peninsula. He termed this the Principle of Equal Unattractiveness, a typical physicist's explanation. Eventually outrageous rents and overcrowding did make the Valley as unattractive as other parts of the United States.

Today the old barn where Shockley started houses a stereo specialty shop. Japanese stereos with Japanese silicon transistors rest on the spot where I sat at my desk, calculating diffusions and working with Shockley on the theory of solar cells. No brass plaque commemorates the birthplace of a new age. In 1963, after ITT nudged Shockley out of his own company, he became a very successful executive consultant for Bell Labs, then held a professorship at Stanford. But William Shockley is known today, not as a scientist of incontestable significance, and not because he understood the potential of the crystal, but because of his controversial opinions on intelligence differences between the races and his shocking proposals for creating an elite group through sperm banks. Sensational reporting of Shockley's views caused the public to turn against him, isolating and estranging the Nobel laureate.

John Bardeen, Shockley's rival and friend from the early days at Bell, became a professor of theoretical physics at the University of Illinois and, in 1972, won a second Nobel Prize. With two young colleagues, John Robert Schrieffer and Leon N. Cooper, Bardeen pieced together the answer to the most difficult puzzle of the solid state: the disappearance of electrical resistance evinced by some metal crystals at very low temperatures, termed superconductivity. That topic then lay dormant for many years until the surprising discovery of new materials was made in 1986.

7

Completing the Circuit

. . . .

BOOM and growth promptly gave the young semiconductor industry its first big headache: too many companies plunged into the transistor business. Every technical advance, every new science-based trick improved yield. Whichever company led the field could cut prices drastically, increase its share of the market, and force the competitors to follow suit. But demand lagged behind the swell of supply; there were just not that many uses for all those devices. Warehouses overflowed with unsold transistors, and financially unstable firms foundered. Some highly educated physicists who headed companies rested on their laurels rather than attending to bookkeeping and day-to-day business routine. Many companies were in over their heads, and their stock turned to waste paper. The early 1960s saw the first crisis.

People began to ask troubling questions. Do we really need so many transistors? Does it make sense to house more and more elements on bigger and bigger silicon crystal wafers? Who is going to buy all these microelectronic devices anyway? The military demand was not nearly as insatiable as had been expected, and it was not possible to sell unlimited numbers of radios and televisions. Computers appeared to be the best market, but the systems were still large, and expensive. At that time most people thought that a large central processor with many time-sharing terminals was the right computer for the business world. The lonely little transistor, a "discrete" individual module, was not at all satisfactory for computers. There was not yet a place for silicon in the memory unit, which still used magnetic beads strung on wires. What was needed was an integrated circuit.

Circuits in Silicon

Both Jack Kilby and his colleagues at Texas Instruments in Dallas and Bob Noyce at Fairchild had the foresight to file patent applications in advance for a system that would house several transistors on a single piece of silicon. The new process also incorporated connections, allowing current to flow from one part of a circuit to another. If these dreams could become a practical reality, the use of crystals would be greatly advanced.

Kilby had experimented with germanium, using thin, bonded gold wires to connect the individual transistors. But in his patent claim Kilby wrote that such connections could also somehow be installed directly on the semiconductor's insulating layer. Texas Instruments announced this finding with a great deal of advertising fanfare. It was the main topic of conversation at the Radio Engineers' Conference in New York in 1959. But Fairchild had already gone further; with their experience with the planar technique, it was only a short step to placing the connection paths directly onto the protective oxide covering.

Jack Kilby and Bob Noyce became involved in a long and bitter legal argument over who was the actual inventor of the integrated circuit and thus entitled to the patent. In the meantime, however, semiconductor technology was maturing. Ever larger teams engaged in research and development on all fronts. Unlike Braun or Laue—or even Shockley, Brattain, and Bardeen—the researchers no longer worked in isolation. A field in which a few individuals had paved the way was turning into the anonymous work of a number of experts. Every form of technology takes a similar route when it leaves the laboratories for the factories.

Computers were still operating with large numbers of identical diodes and transistors connected one after the other. Each logic operation was represented by a series of interconnected transistors that served as a "gate", called TTL, for transistor/transistor-logic circuits, for the ones and zeros. Combining all these gates on one piece of silicon sounded like a good first step. The trick was to integrate a circuit with numerous elements into the silicon lattice.

Everyone was enthusiastic about this idea, but we shared some general reservations about crystals. We were lucky if 50 percent of the transistors on a crystal wafer still functioned after being bombarded in the hot furnace with oxygen and donor atoms. With so many complicated devices, how could a reasonable yield be expected? We assumed that every other transistor would be functional. If each circuit required two transistors, only a fourth—at most—of all the circuits would be functional. If you used three transistors, only an eighth of the circuits would be usable, according to simple probability theory. With four transistors only a sixteenth would be operational.

So the researchers went back to basics, closing in on the crystal structure with X rays and electron microscopes, increasingly exact measuring methods, and more precise calculations. In the sixties, the U.S. Air Force financed the "Physics of Failure" program, which investigated disturbances in the atomic structure of semiconductors. Once again the scientists doing basic research lacked courage; their calculations were pessimistic. The engineers, whose profession was more optimistic, were proven right. They discovered that imperfections in crystals developed only at certain points, mostly at the edge of the heated wafer. Large areas of the wafer remained intact. Circuits could be constructed in these areas because there was enough space for numerous functions, all easily connected internally. Complete circuit diagrams for mathematical logic could now be molded into the interior of the crystal.

One difference between the competing Texans and Californians was that Texas Instruments developed the new circuits with millions of dollars in research contracts for the armed forces. The Air Force needed small, reliable components for its new Minuteman missiles. Most of the breakdowns in their projectiles could be traced back to faulty connections in which wires tore off the semiconductor crystals. Seeing the promise of integrated circuits, the Air Force was willing to pay almost five hundred dollars per circuit. Fairchild, in contrast, valued its technological capabilities so highly that it accepted no help from the weapons industry. At the same time, its entrepreneurs were perfectly willing to sell the military its integrated circuits. The Minuteman gobbled up silicon. With government contracts, Fairchild grew rapidly, soon reaching and then surpassing 100 million dollars in sales.

With the advent of the integrated circuit, semiconducting crystals had truly become intelligent. Logical decision operations were now cheap. It was no longer necessary to establish long leads to a large central processor, make computations and decisions, and send back commands. Logic could be set up directly in every measuring device, at every control point. Instead of being isolated, monolithic monsters, computers could be incorporated into all sorts of electrical devices.

Small Is Beautiful

Production methods in the new electronics industry differed more and more from traditional factory operations. The breakthrough came with the use of semiconductors for memory storage in computers. Several transistors with resistors and charge-storing capacitators could now be imprinted on a single silicon chip. At the edges of these regularly recurring memory sites it was possible to build in the amplifiers and all the switches needed for transferring, reading, and writing in the information. The continual back-and-forth between electricity and magnetism in the memory modules was a thing of the past. The presence or absence of electrical charge caused the device to decide between zero and one. Now only electrical charge was pushed back and forth along short paths. Whatever company first learned to build a memory storage unit would open up the market, for all computers have an insatiable need for memory elements.

In the mid-1960s, the big labs on the East Coast had silicon and its oxide skin completely under control. The simplest possible device had always started with a piece of silicon attached to a metal plate. An electrical charge on the plate would create the opposite charge inside the semiconductor. This simplest of ideas to make a conductor out of a nonconductor, the basis of the first patents, had taken a century to achieve. Electron tubes, the point-contact transistor, and even the junction transistor had been helpful but roundabout ways of reaching this final goal. Crystal surfaces had always prevented realization of the dream, for they held on tight to the electrons, stopped them from flowing and thus rendered switching impossible. But silicon and its oxide layer were a perfectly clean system.

The long-awaited field-effect transistor (Fig. 5) was now possible. The electrical field of the charges introduced to the crystal externally could control a current flowing within the semiconductor, so long as the oxide layers were sufficiently pure. The transistor had found its simplest form. The system was called MOS, for metal-oxide-silicon, in accordance with the order of the layers. The electrical field in the metal penetrates the oxide and switches the current on and off inside the silicon.

Once again the fundamentals were researched by the East Coast giants: IBM, RCA, and Bell Laboratories, with their traditions of surface research. IBM continued to use magnetic storage, although it also researched MOS systems. RCA developed MOS methods in the East, while Fairchild carried out the same work on the West Coast. Not until the end of the sixties did this technology leave the laboratories and head for the semiconductor factories (Fig. 6). Around 1970 an integrated circuit housed, on average, fifty elements; by the late 1980s the number had risen to many hundreds of thousands.

The field-effect transistor gave electronics another big push. On the negative side, the MOS elements were clearly slower because the electrons had to traverse greater distances inside the crystal, and the signals needed more time to reach the gates. In this regard the bipolar transistors definitely had the advantage, and the new system had to work hard to compensate. On the other hand, the MOS systems were simpler and less risky to manufacture. Since the elements could be packed closer together, MOS circuits became increasingly cheaper.

Price Cuts and Learning Curves

In 1970 it cost one penny to store a single yes-or-no decision, known as a "bit," in a piece of silicon; by 1986 this price had fallen to less than a thousandth of a cent. This incredible drop in prices precisely follows the mathematical "law" of microelectronics, which states that every six years, a price will fall to a tenth of the previous price (Fig. 7). Such

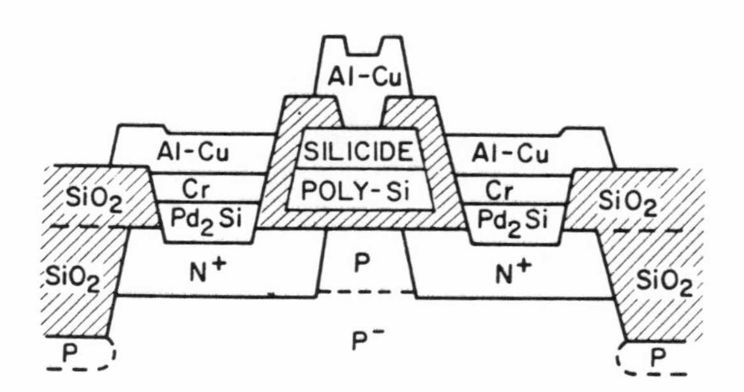

Figure 5. Schematic cross section of a silicon wafer upon which a modern field-effect transistor has been constructed. Note the complexity of the various layers. Current flows from the left through contact layers into the N^+ designated zone of the source. A gate contact consisting of silicides, an aluminum-copper alloy, and polycrystalline silicon can admit or block current in the P^- designated zone. Current leaves the silicon crystal through the drain on the right. The shaded areas consist of the isolating oxide of silicon. The MOS field-effect transistor is one of the most important construction design principles of silicon technology.

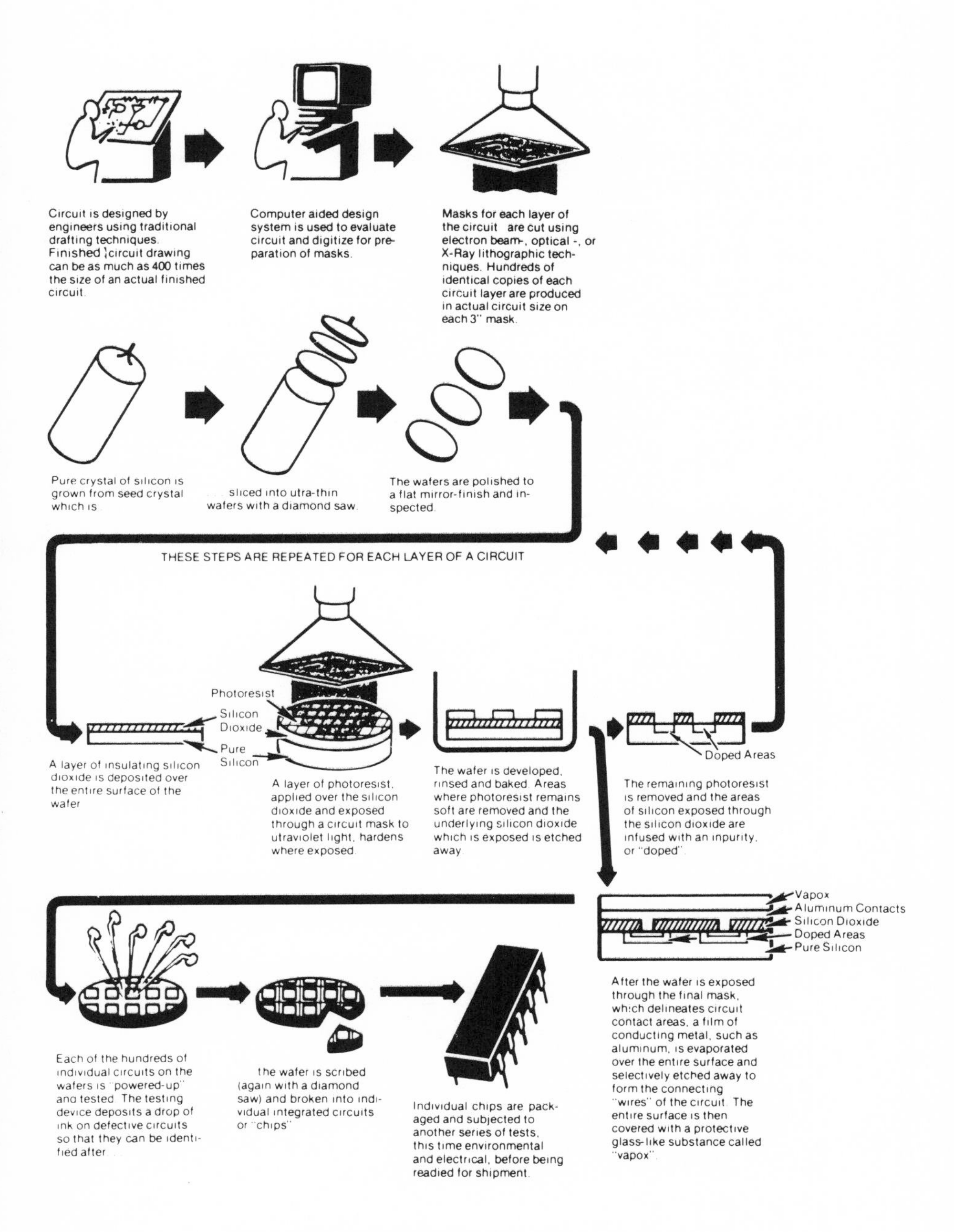

Figure 6. Steps in making an integrated circuit.

drastic cuts in costs were new in technology and explain the increasing pervasiveness of microelectronics in our lives.

Fairchild's Gordon Moore, who had been a Shockley team member, attempted to interpret this law. He argued that these dramatic cost reductions are possible because semiconductor physicists are able to delve deeper and deeper into the crystal's inner structure. Other technologies, such as those in agriculture or metal processing, are limited in their attempts to conserve materials. Agriculture can attain many of its goals by using artificial fertilizers, modern farming equipment, and new breeding methods. But each product—beef or potatoes—does not change much, presenting a definite barrier to further alterations. Microelectronics, however, is able to use fewer and fewer crystal atoms for

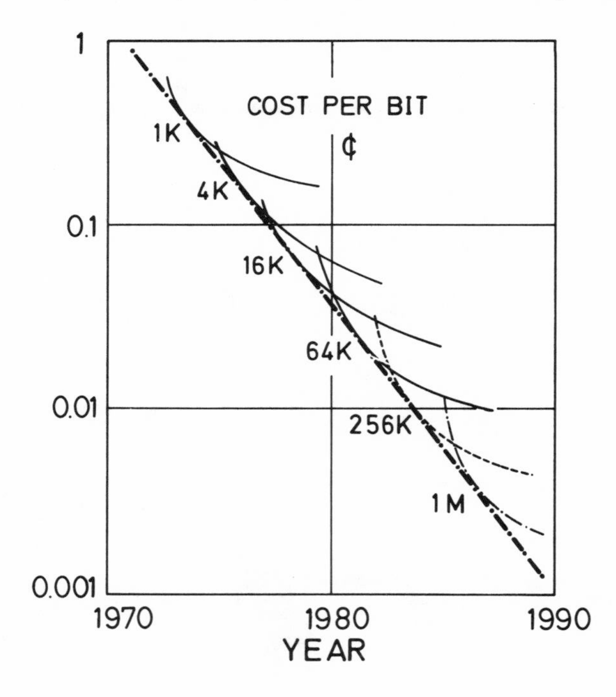

Figure 7. Graph showing Moore's Law of the exponential drop in cost of computer storage function from 1970 to 1990. Costs are per stored information bit, in cents. The individual storage module generations are shown as curves. 1 K means 1,024-bit storage, 4 K denotes four times that amount, and so on. 1 M denotes about 1 million storage cells on a chip. In each generation the cost per unit first drops abruptly, then flattens out. The diagonal dashed line is the tangent, which portrays the exponential drop in costs, since the left axis is not divided linearly but logarithmically, in tenfold jumps.

each storage function. Changes can be made in three main areas. First, the size of the individual transistor can be continually reduced with sufficiently precise instruments and effective masking techniques for transferring a photographic image onto the crystal. Of course, connecting and adapting all the steps to each other requires the greatest care. The surroundings in which the crystal is processed must become increasingly sterile. The finer the structure, the greater the danger of a tiny bit of dust destroying the results.

The second area of savings is in using ever larger silicon crystal wafers that can house more and more chips. Although treating the wafers is expensive, the cost is almost the same for a large wafer as for a small one. The prerequisite in any case is a technique for producing large, flawless crystals.

The third area is the researchers' and engineers' "design cleverness." They are continually making switching operations smaller and more discriminating. "Real estate" is the name circuit technicians give to the space they must parsimoniously utilize. Some specialized firms in Silicon Valley do only one job: they "shrink," or reduce, proposed circuits to two-thirds of their original size. New semiconductor circuits bear no resemblance to those previously manufactured, with their tubes and coarse condensers and resistors. All the strengths and weaknesses of the crystal, comprehended and mastered only after years of dedicated research, are incorporated in the design of the storage cell.

The trends in these three areas reveal why storage costs could fall to less than a thousandth of their original cost in only twenty years. Despite beliefs to the contrary, they will continue to fall. The spreading use of integrated circuits also helped the downward trend. Falling prices created more markets. The computer, a machine that only the government and large corporations could afford to buy and use at the end of World War II, was available even to schoolchildren by the mid-1980s. A manufacturer could calculate what his product would cost if he put it on the market a year or two later. He could also predict when fantastic items would be cheap and readily produced. Following the example of prices in the semiconductor industry, it seemed possible to plan such technological research in the United States—and soon in Japan as well.

Unwilling to rest on their past achievements, the entrepreneurs pushed the limits of research and development. The "learning curve" became the dogma of the Valley. A manufacturer whose storage circuits had one thousand storage cells had to immediately start developing the next generation having four thousand cells. Present profits had to cover this period of research and development. In the beginning, manufacturers had not yet learned just how to work with the new materials. The crystal avenged itself mercilessly: every defect or error led to losses and breakdowns. Only a small percentage of all the chips on the silicon wafer were salable. Yield was the name of the game.

The microelectronics industry had a rule of thumb: even the most complicated circuit will cost no more than five dollars to manufacture if volume is high enough. Each new element is condemned to obsolescence, replaced by a new generation. The exponential drop in prices and its merciless consequences were new.

The restless, dynamic industry underwent phases of frantic activity—night shifts, working weekends, customers clamoring for circuits while the warehouses were empty—followed by periods of massive overproduction and clogged sales channels. There was nothing to bring about a balance. The price war attained unheard-of dimensions yet also led to economic cooperation. A network of small, mutually cooperative, and extremely flexible businesses made Silicon Valley what it is today. Specialized subindustries servicing semiconductors—new tools, ultraclean chemicals, computer programs, measuring and testing instruments—became part of the patchwork. Many of the firms were run by former Fairchild employees.

The neighboring universities also took part in the evolution of the Valley as companies wooed professors away with "advisory" posts. Doctoral candidates worked in the industry at night or during vacations, treating problems from the commercial world of technology as scientific topics. The universities needed experienced professors to teach in the new fields, but had trouble finding them. Industry and academia exchanged the most recent technology, methods, problems, and ideas.

The pressure of prices and market conditions resulted in the internationalization of labor. Anyone entering a typical semiconductor factory by the back door will find big cartons of silicon wafers labeled with destinations halfway around the world—Korea, the Philippines, Taiwan, Hong Kong, Malaysia. In these offshore production facilities, the silicon wafers are scribed with diamonds, then broken apart into individual chips and processed by workers who are paid rock-bottom local wages. After the chip is mounted on a ceramic or plastic base, gold wires as thin as a human hair connect the chip to its metallic contacts with the outside world. The next step is the intensive testing of all the chip's functions. After the chips have been checked, sorted, and hermetically packaged, they are sent back to America. Californians, who enjoy some of the highest wages in the United States, can afford only to plan, design, research, and develop the processing; the production work has to be done abroad. The secrets of the circuits and the art of treating the entire silicon wafer remained in the Valley.

When silicon wafers were sent to East Asia for processing, native experts soon appeared. Eager for a bigger piece of the pie, nations such as Singapore directed their educational policies and economic strategies toward more than just assembling and testing chips for wages; they started their own firms, manufacturing not only chips, but entire subsystems and even complete electronic instruments. By 1982 the Philippines' leading exports were electronics parts. Free trade zones with significant duty exemptions aided export to American customers.

American companies have always set up manufacturing outposts in other countries, especially in Europe, to quickly serve those markets. The labor intensiveness of work with semiconductor crystals led to very intense "offshore" assembly and testing in Europe and, even more important, in Third World countries. A survey in 1974 showed that U.S. companies maintained twice as many offshore plants in the developing countries as in developed countries. The Far East and Mexico ranked high on the list of locations. The governments of these countries subsidized the construction of assembly plants and granted tax and duty favors.

The outflow of jobs was not greeted with enthusiasm in the United States. Two items on the United States Tariff Schedule specifically allowed American companies to reimport their processed chips after mounting and testing either completely duty-free or merely with a duty on the value added offshore. The rapid drop of semiconductor circuit prices was accelerated by this policy.

Japan's policy was exactly the opposite: automation at home and attention to chip yield. The Japanese government did not allow duty-free reimport. The American offshore tariff laws were the subject of continued political discussion for over a decade; in the eighties the relative significance of labor costs decreased in relation to the costs of sophisticated and automated semiconductor machine tools.

But it was a technical component that changed the prospects for jobs in the semiconductor industry in California. As automatic packaging plants and testing machines became faster and more sophisticated, a new machine tool industry was created. The silicon in the circuits controlling these new machines became the most important aid in manufacturing new silicon products. Under the pressure of this development, the jobs that had gone to East Asia returned to expensive California.

Western Europeans were less interested in new impulses from the Valley; they rarely visited the laboratories. The only German industrial expert I saw in the Valley was on a tourist excursion. Of course, Germans were enjoying the prosperity of their postwar "economic miracle," based on conventional forms of manufacturing. Yet Charles de Gaulle, an early visitor to Silicon Valley, delivered an insightful speech at Stanford. And in the early seventies the French industrial firm Schlumberger bought the Fairchild company, at a time when most of the founders had already left to establish their own businesses. Nevertheless the purchase came as a sad shock to Valley managers. By then the Valley was losing its independence, its happy-go-lucky approach, its unconventional style. The loss was the price paid for international recognition. This was only the beginning; Asians and, finally, Europeans began to snap up even the less active semiconductor firms. The cause of this new rush to the Valley? The microprocessor!

Microprocessors

If logic and storage can be built into a silicon crystal, most likely the two parts can be united, and a complete computer housed within a semiconductor. Or so the researchers thought. In the early stages each integrated circuit was custom-designed. A silicon firm would be given a draft of a circuit, then photo masks would be produced to create the desired combination on the piece of silicon.

Noyce's company, Intel, which was dedicated to the MOS technique, at one point was saddled with a difficult custom job. The engineers thought about finding more than one customer for the same piece of equipment. They realized that in modular form, chips could be multipurpose products. Chips had become so cheap that they were being used in many more applications. Combined in modules, chips could be used for calculating, data acquisition, control, regulation, and monitoring. Complete industrial processors, previously composed of individual components on a circuit board, could now be placed in the microscopically small piece of silicon called the microprocessor. This new term was to become the Valley's new catchword.

The first microprocessors could handle only four bits of digitized information. The breakthrough came with machines that could process eight bits, which was sufficient for quite a number of applications. Systems created from building blocks of interlocking chips became increasingly varied. In the middle of the system was the central processor, actually a complete computer, less than an eighth of an inch square, which used to take up an entire room. The processor, operating on just five volts of power and timed by a tiny quartz crystal, could add and multiply. It was combined with memories, some of which functioned like scratchpads: data could be read in, scanned, erased, or retained. Other memories had operating instructions for the functional sequence of calculations, operations, commands, and information programmed into the silicon. And other chips could send and receive information for adapting to measuring instruments or the telephone network. Converters transformed data into the digital form necessary for computer processing.

When these inexpensive, multipurpose building blocks were combined into specialized programmable modules, new markets opened up, and Silicon Valley underwent another growth surge. Before, the silicon industry had served as a supplier of transistors for electronic instruments. People selling transistors enjoyed about as much prestige as suppliers of screws, lightbulbs, or cable. But now this situation changed radically. Firms that previously had manufactured only components were selling a variety of adaptable computers and complete systems.

The entrepreneurs recruited young experts who were knowledgeable in both computer construction and programming. This challenge spurred the development of new circuit principles, designs for faster operation, and, of course, new kinds of programs. Technological mastery of the crystal remained important, but took its place next to comprehension of entire systems.

New conflicts arose. Suppliers of individual parts began to compete fiercely with the large firms. Transistor firms aimed for "forward integration"—building whole computers and other products. Why not manufacture watches or even toys? But the equipment industry struck back with the threat of backward integration. In order to compete, conventional companies that manufactured a product such as cash registers or scales had to consider establishing subsidiary plants to produce integrated circuits. Each Valley business worked hard to convince buyers that no other company could compare with it in terms of speed, creativity, and, of course, price. "Everything you make with springs, screws, and gears, you can make better with microprocessors" became their slogan. And so the world began to court the California silicon specialists. European firms bought partial or total shares of Valley businesses, and East Coast magnates attempted to get a foothold in the Valley and access to its technology.

In the early 1950s about a hundred computer systems were sold each year in the United States; by 1960 the number was some two thousand. In 1968 the figure had risen to fifteen thousand—and in 1975 a quarter of a million computers were sold worldwide. No one had thought that an initially very expensive and large system would so rapidly become an everyday product. At the same time, rock-bottom prices of microprocessor components enabled Valley businesses to expand into a variety of consumer products—watches, games, office equipment, and the whole scope of radio and television electronics. While most firms in Silicon Valley had initially hedged their bets on the military market, the percentage of sales to the military shrank dramatically. Silicon Valley evolved with the support of, but not dependence on, weapons technology.

One evening I sat chatting with friends in the Intel cafeteria. Although it was late, many of the young engineers were still at work. One of them wearily walked up to our table, sat down with his Coke, and sighed. "You do your best, rack your brain, use the most sophisticated mathematics—and to what end? Our customer uses the memories and chips for the dumbest and most primitive games. Have you heard about the latest video games? You have to run over people, and then the thing peeps and flashes and gives you bonus points. Good Lord, we're working with the highest form of technology for this kind of garbage." His sentiment was justified. The public's addiction to electronic games had swelled out of proportion. Video games companies bought up to half a billion dollars' worth of semiconductor devices annually. The abrupt and hectic growth resulted in some big headaches and bankruptcies a few years later, but overall it stabilized the Valley.

The microprocessor further changed the peninsula near San Francisco. Gone were the country roads, the little gardens and fruit groves. The communities grew rich; money poured into schools and libraries. Large hotels and high-rise office buildings appeared. One could find first-class bookshops and fine European wines around Palo Alto. This area, with some of the highest incomes in the country, began to evince a touch of cultural snobbery.

Yet the risky, high-growth business brought disruption as well. At the bar in Rickey's Hotel, the standard sarcastic greeting was: "Well, are you still divorced from the same wife, or is it a new one?" Step-parents seemed to be the norm for schoolchildren. Drugs, a problem everywhere in the Western world, were particularly pervasive. Valley people developed their own black-humor term to describe their lifestyle: "burn-out," the last phase, just like the end of an overloaded circuit.

The happy-go-lucky, cheeky, we-can-do-it approach gave way to another style. Foreign companies hired brains from the original companies and destroyed the old cliques. Property was hard to come by; around the late seventies, only ten percent of the land remained undeveloped. Skyrocketing prices for housing and property restricted further growth of Valley businesses. The reliance on manufacturing gave way to intensive creation of value, particularly through the design and development of software. The businesses spread out to Utah, Idaho, Oregon, and Nevada, which now share in the wealth.

Concerned about environmental protection, Democratic administrations in Washington and Sacramento initiated new regulations. Though it is one of the cleanest industries, silicon has contributed to air and water pollution. Fluorine compounds and the solvents for photo resists are among the culprits. Concern arose for the health of workers encapsulated in protective gear, working long shifts in a secluded, sterile environment under constant yellow illumination. Some environmental regulations went too far; Bob Noyce told me that in one case substances filtered out of tap water could not be added to waste water but had to be carted away in containers. Laws like this forced many silicon firms to search for areas with more lenient laws.

Silicon Valley lost its tranquility. Stories of industrial espionage abounded. As industrial spies from the Eastern Bloc and the Far East swooped down on the region looking for valuable high technology secrets, the FBI laid traps. Nobody could afford open doors or casual conversations. Colleagues who went separate ways to establish new firms often met again in court. Those who made off with confidential business information were often prosecuted, and lawyers, patent attorneys, security police, and informants became important figures. The disdained characteristics of the traditional, stiff industrial structure crept into the Valley. California's semiconductor people used to claim: "We create wealth, everyone else merely distributes it!" This proud statement became less and less convincing.

Then the Valley also lost its leading position in the world. A dramatic upsurge in business in 1984 was followed by an unusually severe downturn of the economy in computers and electronics the following year. American firms were losing ground as Japanese companies moved into the top spots. Personnel were laid off, entire factories mothballed, and construction of the new "fabs," as the silicon wafer fabrication plants are called, was halted. A new war broke out across the Pacific, what the Japanese call *nichi-bei handotai senso*—the Japanese-American semiconductor war!

Guglielmo Marconi, considered the father of wireless telegraphy.

Ferdinand Braun in his physics laboratory at the University of Strasbourg.

Lee De Forest,
an inventor of the amplifier.

Wilhelm Conrad Roentgen,
the discoverer of X rays.

Max von Laue as a student at Strasbourg.

Apparatus used by Laue, Knipping, and Friedrich in 1912 to study the interference of X rays in crystals.

Postage stamp honoring Max von Laue's discoveries.

Robert Wichard Pohl, experimental physicist and crystal researcher at the University of Göttingen.

Walter Schottky, who in 1939 explained the electronic processes occurring near the semiconductor boundary layer. Many microelectronic terms, including the Schottky diode, were named for him.

Aerial view of Bell Laboratories, Murray Hill, New Jersey.

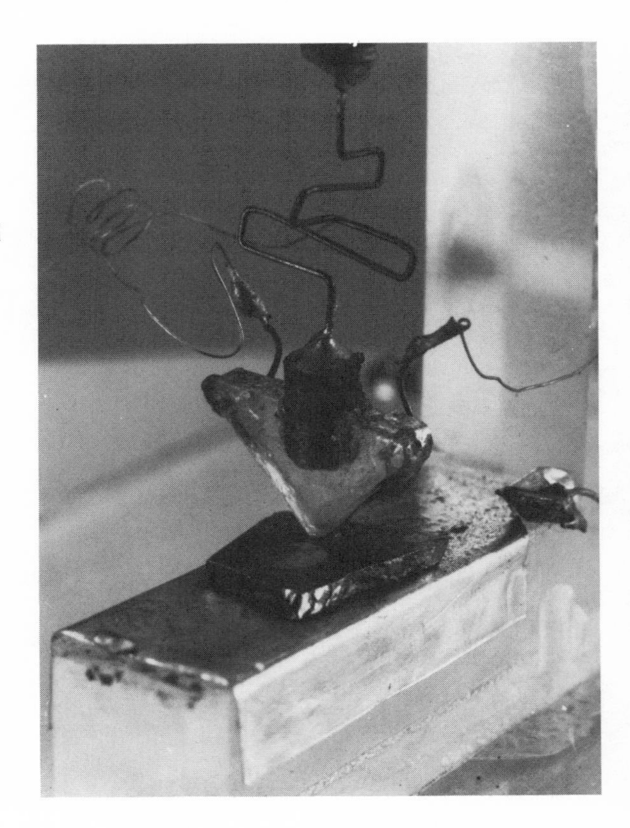

The first point contact transistor was assembled at Bell Laboratories on December 23, 1947. The plastic triangle with the current leads presses down on a slice of germanium.

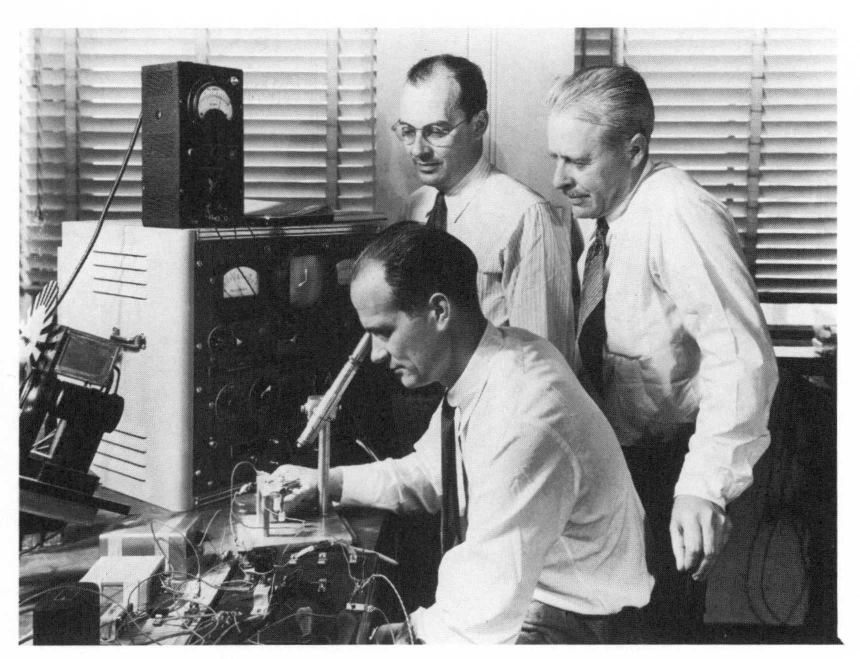

The inventors of the transistor, Nobel Prize winner William Shockley (seated), John Bardeen (left), and Walter H. Brattain (right).

Konrad Zuse, born in 1910, a pioneer in the development of computers.

Robert Noyce, who left Shockley to start a new company.

The "Shockley Eight" (Robert Noyce is standing, center, with glass) and others toast William Shockley (seated at end of table) on the day in 1956 when he was awarded the Nobel Prize in physics.

Jack Kilby, one of the inventors of the integrated circuit.

A field effect transistor.

The Intel 4004 microprocessor.

Leo Esaki of IBM, who received a Nobel Prize in 1973.

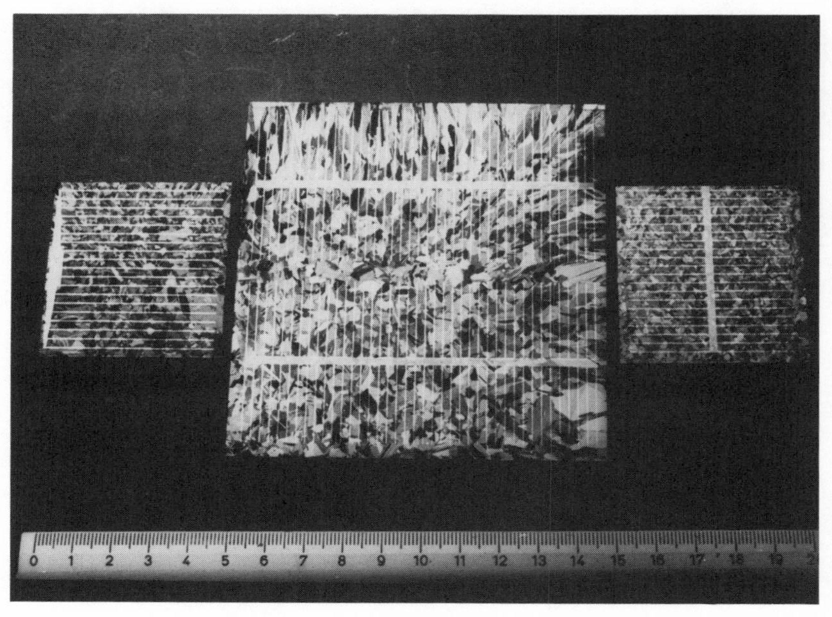

Silicon solar cells.

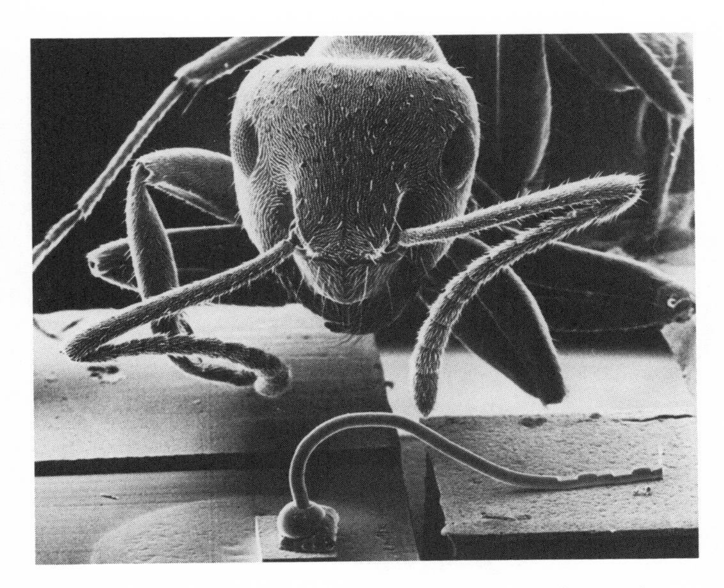

A gallium arsenide laser diode in front of an ant, to show scale. Current flows through the wire and, via a small metal contact, into the metal plate below. Light, generated directly by the current, is emitted at the carefully prepared cleavage face of the crystal (below, middle). The light can then be guided into a glass fiber for communications transmissions. The photo was made with a scanning electron microscope.

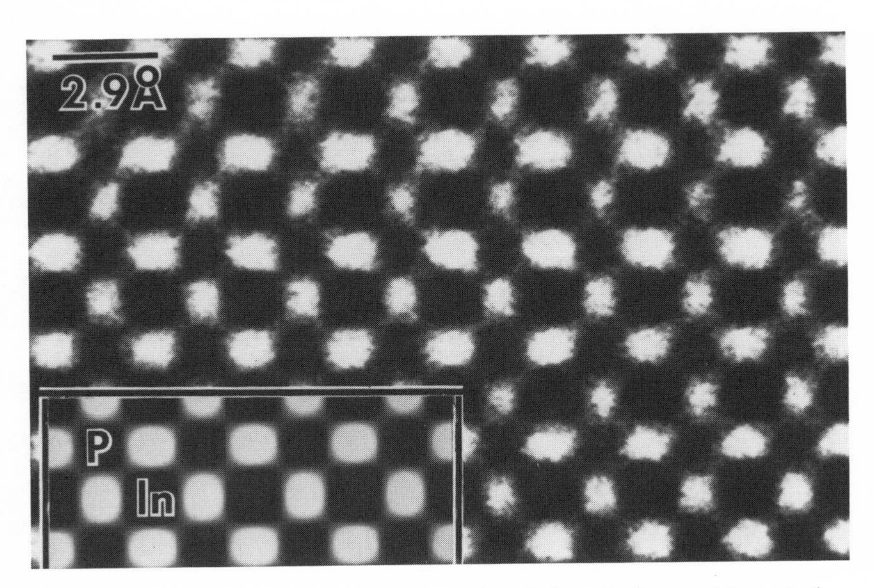

Lattice image of the semiconductor indium phosphide, which is used in optical communications. This electron microscope photograph shows individual columns of atoms (black squares). The indium atoms are larger than the phosphorus atoms. The inset shows a computer simulation of the same structure.

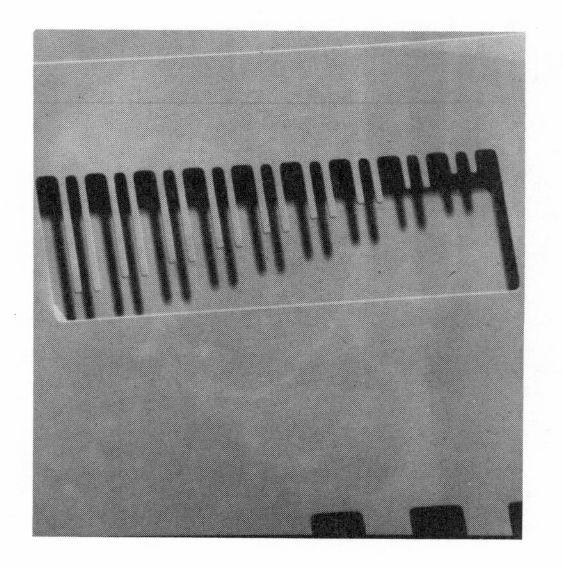

Rods, one one-hundredth of a millimeter wide, etched out of silicon. The rods can oscillate.

The ink jet nozzle of a computer printer. The nozzle is 50 micrometers in diameter, with a wall thickness of 3 micrometers.

8
Handotai Senso

. . . .

In Japan, everything runs like clockwork. When I was visiting the country in 1980, my Japanese friends showed me their latest research. At the door to the research lab of one industrial corporation, I obediently removed my shoes, put on wooden sandals, and padded through the modestly furnished laboratories, workshops, and lecture halls. The small rooms housed intense activity. My hosts fired precise questions, always under the cloak of etiquette. It was something of a shock to learn how well informed they were about the state of technology in the United States and Europe.

Nihon-ichi

The old taxi driver who drove me to my afternoon appointment at a competitor's company struck up a conversation. In halting English he asked, did I know anything about the new computers and semiconductors? I had to, he reasoned, or why would I be visiting two such labs? Could I perhaps answer a question that had been bothering him since he had heard the news that morning? The day before, IBM in New York had reported that it had produced a device that could store 64,000 bits of information. He was just a taxi driver, he said, and did not understand such things, but he knew that the Japanese simply had to catch up with IBM by 1985. Would IBM's announcement ruin their chances? I assured him that it was still just a pilot project from the United States and that Japan would most likely achieve its goal. But my taxi driver remained worried; he felt personally affected. A week later in Frankfurt, my German taxi driver was also concerned—but his worry was that his favorite soccer team was not doing well! He had never heard of silicon.

A friend once gave me a little samurai doll bearing a paper scroll on which was printed *Nihon-ichi*: Japan Number One. Japan wants to be first in auto exports, steel production, and, of course, silicon. They aim to win the battle for control of the silicon crystal and the microelectronics industry. By 1983 Japanese semiconductor companies controlled almost three-fourths of the international free market for silicon chips that stored 64 kilobits of information, and by 1986 they almost completely dominated the 1-megabit memory generation.

People who travel to East Asia are shocked to see how well our competitors have learned to work with semiconductor crystals. Intelligence, discipline, punctuality, and cleanliness, as well as servility and self-sacrifice, are the qualities the Japanese offer the semiconductor industry. A country with scarce raw materials, Japan has dedicated itself to conquering new international markets. And it has succeeded.

When World War II ended in 1945, roughly half of all Japanese worked in agriculture and fishing; not even a quarter were employed in industry. After Japan's defeat the rest of the world assumed that this island was at best only slightly more advanced than the other developing countries. Japan accounted for barely 3 percent of world production. Its main value to the United States was as a cornerstone for American foreign policy in the Pacific. To prevent the communists from gaining a foothold among Japanese workers after the war, General Douglas MacArthur pushed through a plan in which industry guaranteed each worker a secure, lifetime job. Job security helped engender a positive sense of community among Japanese workers, who often spend their entire career within one company. By the late 1980s the percentage of the workforce in agriculture had fallen below 10 percent, a figure common to most of the industrial nations.

After World War II the Japanese struggled to overcome their reputation for producing cheap, shoddy goods and imitating Western ideas. The American military paid high prices for goods and in return demanded dependability and exactitude. Government agencies took on the role of ensuring high quality. The Japanese had to counter the accusation of miserable quality with the very opposite: precision. The British occupying powers had once required German products to be stamped "Made in Germany" to warn the customer of their bad quality. In time this stamp became a seal of quality! History repeated itself with the "Made in Japan" label.

In the early postwar years Japan could not rebuild, as Europe did; it had to start from the ground up. The economic miracle of the Federal Republic of Germany, achieved by massive production and export of well-known technology, found no counterpart in Japan. The Japanese had to conquer markets with novel products and new technologies. The Japanese, though, learned a lot from the German experience. They first built up their heavy industries, especially steel and shipbuilding, using the most modern technology. From the outset Japan planned to take competitive advantage of more than just its cheap labor. Establishing contacts with the United States and Europe, the government promoted the importation of science and know-how. Legions of young Japanese were sent abroad to work and study, and many saw at first hand the earliest developments at Bell Labs and in Silicon Valley. Back home in Tokyo, the question was not how to defend the old technologies but how to plan for the future. Japanese culture emphasizes work for the common good. Every citizen is reminded that his densely populated country, with few energy resources, must conquer the international market with the newest products of the highest quality—or starve.

The Western researcher's freedom to choose his tasks and methods is unknown in Japan. For many years the Japanese physicists had to watch enviously, as specialists in the United States and Europe had easy access to computers, measuring equipment, and other tools to do first-class, interesting basic research. Japan, however, forbids this easy route of importing just for the sake of intellectually exciting research. Researchers must first understand, then copy and fabricate the equipment. The scientific community therefore underestimated until well into the sixties what was happening in Japan. By the time the world understood the extent of Japanese know-how, the Japanese had already created their own technological infrastructure and did not have to rely on imports.

Japan has used every means available to protect itself from technology imports. Almost none of the leading American and European science and technology firms have gained a foothold in the country. They have found it nearly impossible to establish affiliates or obtain a controlling interest in Japanese firms. In general, the Japanese allow cooperative efforts and joint ventures but retain the final say in decision making. Adding to Japan's independent stance is the government's strong network of support for technology and research. Even small businesses are eligible for tax cuts and grants for applied research.

Western researchers are schooled in skepticism, self-awareness, and independent work. They are accustomed to questioning the results of their colleagues' work. If they publish their own results in scientific journals, they expect in return the rewards of esteem or a higher position. The cult of individualism has led to great successes in Western physics research, yet that competitive spirit can erect barriers between the universities, the factories, and the research laboratories. Japan rejects this aspect of Western individualism. A young Japanese engineer will not discriminate against a result merely because it is not his. Information is valued as a material resource.

The emperor lends his prestige and pays 50,000 yen out of his own pocket to Japan's Institute for Invention and Innovation. That donation, a very small contribution to the budget of 12 billion yen, signifies the importance granted to the organization. Promoting and recognizing inventors and creating a friendly atmosphere for technology and industry are the tasks of the Hatsumei Kyokai Institute, which pays particular attention to young people. Annual science fairs for young inventors may receive up to 100,000 entries, and the best projects receive prizes under the auspices of the emperor. The Hatsumei Kyokai also sponsors inventors' clubs, which children can join beginning in the third grade, if they qualify. Members receive guidance and the use of a remarkable collection of equipment. The most important function of this institute is to disseminate technical knowledge and patent specifications to the whole country.

Since the steel, petrochemical, aluminum, paper, and synthetic fiber industries are heavily dependent on the importation of costly raw materials, the government has attempted to reduce their share of the economy. The relatively weak position of these industries has become even weaker since the oil crisis of the 1970s, but they have been replaced by microelectronics. The challenge of housing more and more functions in smaller and smaller semiconductor crystals and the creation of new products became the new Japanese battle cry. By the end of the 1980s the goal was to catch up with—and surpass—IBM. Even my Japanese taxi driver was aware of the challenge.

Chips for the Rising Sun

All Japanese financial statistics emphasize their success in semiconductor devices. In 1970 Japan still had to import a third of its integrated circuits from Silicon Valley, Texas, and Arizona. By 1982, though, the tables were turned: Japan exported to the United States three times as many circuits as America exported to Japan, achieving a trade surplus of 200 million dollars. For the entire sector of semiconductor products, the trade surplus had grown to more than 500 million dollars—in just one year! Silicon chips have become such an important commodity that the Japanese financial papers now furnish daily quotations of the fluctuating price of 256 kbit memory chips, right along with the prices of gold, copper, and tin.

The situation for U.S. firms rapidly deteriorated in 1985 and 1986. Texas Instruments, for many years the world leader in sales of integrated circuits, lost that position to the Japanese company NEC. In jumping from third to first place, NEC also surpassed the Arizona-based Motorola Company. The disastrous year of 1985 reduced semiconductor revenues for the American companies by a whopping 27 percent, while it reduced the Japanese sales volume by only 5.8 percent. Since 1980 the Japanese have consistently increased their worldwide market share by almost 3 percent each year; most Japanese leaders now expect to beat the United States as early as 1988 and certainly by 1990.

How did the Japanese penetrate the field of semiconductor electronics? What instruments did they use to so brilliantly resolve the difficulties? The raw materials used in the semiconductor war were easy to import. Unlike iron ore or concrete, they weighed very little. Moreover, information on the new technology could be had for the asking.

A Japanese who mastered English could tap into additional Western sources. Foreigners say the Japanese are better at organizing information flow than at creating new information, but this remark does not disturb Nippon. For decades, Japanese experts have been combing Western laboratories and universities. When he was a student, Makoto Kikuchi, now the head of research at Sony, was sent to MIT to study electronics. Later he attended all of the important conferences in the United States. One day, while sitting with his American colleagues in the cafeteria at Bell Labs, he asked them how work was progressing on the new "functional devices." Stunned, his American friends asked: "What on earth are functional devices?" Now it was the Japanese scientist's turn to be surprised: "Your own Dr. Morton wrote an article about them! Everybody in Japan is talking about nothing else. You mean to say you've never even heard of them?" The Japanese had taken even the slightest hints and speculations much more seriously than Morton's colleagues and staff members! Kikuchi never tired of quoting this conversation, using it as evidence of the Americans' oversight.

Japanese experts get extremely irritated when insulted Western researchers accuse the Japanese of having robbed them. Americans complain that Japan has contributed little to basic research in this new field, instead using American research to conquer American markets. But in Japan research is approached more soberly and systematically than it is in the West. The Japanese have set rules, even about the classification of research work. They do not consider basic research and applications as opposites. In their terminology the opposite of "basic" is not "applied," but simply "not basic"; the opposite of "applied" would be something like "not applicable."

Research is divided into four categories. The first is basic research that is not to be applied, which Westerners would term pure research. The second is basic research seeking new principles and methods that can be applied to technological and social problems. The third is developmental work to create new practical applications of scientific knowledge. The fourth category, comprising nonbasic development that makes no new contribution, is considered lacking in orientation as well as scientific exactitude and is avoided.

Rather than disdain applied research, the Japanese reward it. One of the largest electronics firms, Matsushita, sponsors the Japan Prize, worth the unusually high sum of 50 million yen, which is intended to compete with the Nobel Prize. But while the Nobel Prize rewards basic research, the Japan Prize deliberately recognizes applicable basic research.

Before the turn of the century, in imitation of European countries, imperial Japan modernized through its prestigious, highly selective universities. The engineering sciences gained in stature during that phase. Influenced by the technological progress of the Western powers, the Japanese consciously emphasized applicable research from the very beginning. Leading the race for development in this area was engineering, but chemistry, synthetics, textiles, mechanics, and, especially, electronics also helped develop the modern infrastructure. Military ambition played a role in accelerating the pace of modernization.

Up until World War II the Japanese often looked to European models of development, especially those of Germany and England. After the war, however, the United States became the example. Almost from the beginning the semiconductor crystal held the spotlight; microelectronics was recognized early as a very important area. In fact, the Japanese probably realized the importance of California's technological phenomenon before the Americans did.

The Japanese guided their microelectronics sector to replace older industries. A strong domestic market for electronic entertainment products strengthened the trend and was probably more important than direct government support. When accused of unfair competition because of government support, the Japanese counter by claiming that only a small part of Japanese research is financed directly by the state, that competition among the individual high-tech firms is what cuts prices and increases productivity. The *Journal of Japanese Trade and Industry* published the following comparisons of the government's share in total research expenditures (including defense): France, 58 percent; Great Britain, 48 percent; United States, 48 percent; Federal Republic of Germany, 44 percent; Japan, 28 percent. The two nations that lost World War II are naturally weaker in the armaments industry. But even if defense expenditures are disregarded, Japan still brings up the rear in government expenditures on research: France, 47 percent; Federal Republic of Germany, 41 percent; United States, 33 percent; Great Britain 32 percent; and Japan, only 28 percent. It is clear that direct state intervention, like that in France, especially, is in no way the main instrument promoting technology in Japan.

MITI Takes the Reins

Direct government financing of technological development is not so necessary in a country that enjoys a national consensus; government offices serve more to guide efforts and formulate common goals. One of the most important Japanese institutions serving this purpose is the Tsusansho, the Ministry of International Trade and Industry, known as MITI. These four letters often frighten American competitors, who assume they stand for an omnipotent superministry, a formidable opponent that unites economic and foreign policy. But MITI is not so formidable.

During the war MITI coordinated much of the Japanese economy, but it relinquished this function immediately after Japan's surrender. Not until the 1950s, about when Japan was consciously accelerating its industrial research and development, did MITI begin to broaden its scope. At that time the most serious problems were technology imports and scarce dollars. MITI allocated foreign currencies and thus supervised investments. All licensing contracts with foreign partners, as well as plans for producing and exporting products, landed on the desks of

MITI staff, who were leading experts, knowledgeable in technology, and with social prestige.

Japan carried no significant weight in international politics after the war and had no military power elite; the country compensated with its economic policies. The young Federal Republic of Germany experienced the same situation, but its free market economy would not have permitted a government agency like MITI to have such power in technological planning.

MITI has four main tasks. The first is to foster trade, which is critical to Japan's survival. The country has to fight hard against protective tariffs and other restrictions on imports of Japanese goods. The second task is to monitor market trends, make technical information available to industry, and make sure it gets incorporated into a common strategy. To this end MITI has established a number of consultative committees, often chaired by its own people to ensure consensus, even though they do not have direct authority. The third task is to improve international cooperation, alleviating tensions over trade imbalances and allowing Japan to take part in international projects. Finally, MITI helps regulate foreign trade, promoting exports and monitoring imports. In 1958 JETRO, an organization patterned after traditional British models, was established to promote foreign commerce. JETRO has become an important factor in Japan's presence abroad by using the diplomatic corps to gather trade information.

According to the statistics, the Japanese government directly finances only a small part of silicon research and development. But MITI is doing its part, influencing interest rates and credit terms and thus fortifying against foreign competitors.

In Japan MITI oversees agencies dealing with economics, business law, and patent law as well as technology and science laboratories. One research lab, the Electrotechnical Laboratory, is modeled on the National Bureau of Standards in Washington and the British National Physical Laboratory in Teddington. Originally it tested electrical equipment and measured and calibrated electrical standards. But in the 1950s, soon after the transistor was invented, the lab began conducting its own research. This relatively neutral organization easily made contacts with the world's great research facilities, and it became an important source of excellently trained personnel for Japanese firms. Wherever possible, industries hired the laboratory's specialists to direct their research labs.

MITI's strategy of using "neutral" research labs to foster commercial projects became painfully apparent to American observers, not so much by fancy basic research results and Nobel Prizes in physics and chemistry, but by their success in the global marketplace. American observers agree that Japan has an excellent technology policy, but no research policy, exactly the reverse of the American situation. America has excellent support from the government for research, but no coherent strategy and support for technology and trade.

All Japanese firms attempt to attract MITI specialists, for this means a direct connection to the most powerful regulatory body in Japan. A company cannot achieve anything without the help and blessing of MITI. Access to information and data, early acceptance of plans, and a say in decision making—all of these require going through MITI. But the Japanese do not permit MITI to dictate economic policy. Japan is successful in settling political disagreements and conflicting economic interests for a number of reasons. First, the West's polarization between research and economic policies, as intensive as it is fruitless, is largely absent in Japan. Second, industrial companies that have a dispute often turn to MITI as an arbiter. And Japanese culture and organizational structures promote harmony. Most businesses are based on the traditions of agreement and avoidance of open conflicts and loss of face. The oldest committee member at any negotiation automatically has the most authority, and often the MITI delegate is the most senior person.

MITI calls its far-reaching programs for Japan's future technological development "visions." But these programs serve as a framework for discussion and planning, not a rigid economic scheme, as in the planned economies of the eastern bloc nations. MITI does, however, place strict, sometimes unpopular, limits on less profitable, desirable, or environmentally questionable industries.

MITI's best-known vision concerned the crystal. When I was living in California I got to know a rather shy and uncertain Japanese scientist, Yasuo Tarui, who had been sent to Stanford from the Electrotechnical Laboratory in Tokyo. Before returning to Japan, he surveyed all of the developments in Silicon Valley. I kept in touch with him and learned a few years later that the Electrotechnical Laboratory had delegated him to head a joint project of Japanese industry, the universities, and the government: the VLSI Program, or Very Large Scale Integration of semiconductor devices, one of the very powerful projects that finally secured Japanese domination in silicon memory chips.

Having gathered the necessary information, the Japanese began their VLSI program—an ambitious effort to put many components onto large circuits. (They did not, however, merge all of the many semiconductor firms into one supercompany.) The government and industries cooperated on research and planning, and consulted on required norms and preliminary tests. After the start-up phase, MITI and all other state agencies were to withdraw from the program, although few observers in the United States believed that the government would withdraw. Americans worried that the VLSI program was simply the first step toward a mammoth "Japan, Incorporated" structure.

But the rigorous program did operate according to plan. Teams from individual firms were placed in two large groups. Through competition and through division of labor, they worked out all of the prerequisites. The teams developed new photo resist methods for transferring fine patterns onto silicon. They invented new machines, tested the prototypes, and passed them on to the manufacturing plants, which, in turn, created new products. MITI's plan guaranteed that Japanese manufacturers would use the new precision tools for silicon crystal processing. To conquer the world market, their new machines had to produce circuits at least comparable in quality to those of the American competition.

The silicon crystal has demanded increasingly sophisticated technology. Lasers, used to regulate distances of less than a thousandth of a millimeter, have become more and more refined. Ions formed in large stainless-steel furnaces are used to etch the crystal's surface. The temperatures in the furnaces where the foreign atoms penetrate the crystal have to be regulated to within a fraction of a degree. Machines gently and cleanly transfer the brittle silicon chips from one processing step to the next.

Unlike those in the United States, the researchers and technicians in Japan do not compete or duplicate their efforts. Through cooperation new technology is available to all the members of the development team, enabling a work output higher than the West could ever imagine. If the researchers need scientific assistance, the universities lend a hand. The giant scope of testing, the stubborn determination to measure even the slightest change in characteristic values, annoys many Westerners. The West's notion of discovering a principle through a few intelligent experiments does not satisfy the Japanese; their mass production is based on mass experiments.

In one VLSI materials lab I visited during a stay in Japan, thousands and thousands of silicon wafers coursed through furnaces and machines. Heading the lab group was a member of MITI's Electrotechnical Laboratory, a man who had been sent to the United States years before. At Bell he had learned the techniques of breeding and monitoring crystals, and at home he put this knowledge into practice. The lab could discover every weakness in the crystal structure. The researchers tested every imaginable combination of manufacturing processes and investigated every type of foreign admixtures in the silicon crystal.

From the podium at the International Conference on the Physics of Semiconductors, held in 1980 in Kyoto, I watched the Western experts in the audience react to descriptions of this approach to materials. Upon hearing that Japanese competitors had not been able to uncover the principle with just a few directed experiments, the Westerners at first grinned. The Japanese, they assumed, were forced by ignorance and inexperience to conduct thousands of experiments on every conceivable possibility. In the course of the presentation, however, the breadth and depth of the hosts' work became obvious. Their wealth of knowledge and experience began to make the Western guests anxious. Their faces, initially so confident, revealed worry.

A few years later Nippon cornered almost the entire world market for the newest silicon memory chips. Their processes were cleaner and more reliable than those in the United States. Moreover, they knew about all the defects that could interfere with the silicon crystal's functioning, and they knew how to increase yield dramatically. By 1983 four of the world's ten largest semiconductor manufacturers were Japanese. Only one European firm, Philips, was still in this league, and within four years it had fallen to sixth place! In 1985 a Japanese company became number one.

The Japanese also made great progress in telephone systems, which are really nothing more than big computers that find the fastest connection for putting a call through. The exchanges, which used to use mechanical switches, now function electronically. The range of information transmitted telephonically is vast; speech and data flow, and television pictures and concert-quality music are digitally encoded, transmitted, then dissected and passed on to the receiver. Shockley's and Kelly's dream of a fully electronic solid-state telephone is a reality today.

Nippon has provided itself with a high density of telephone connections within a short time. Bell Telephone Laboratories is imitated in and around Tokyo with great success. The powerful Nippon Telegraph and Telephone, or NTT, has given rise to major Japanese supply companies that are making inroads into the telephone market worldwide. Initially a government agency, NTT reinvested all its profits in research and development. It does not manufacture any telephones, but merely installs them, as do the German Federal Postal Services and many other telephone agencies. Nevertheless, NTT has five large research and development laboratories. In Atsugi, close to sacred Mount Fuji, at a crystal research center, the researchers are current on the exact state of the art in modern technology in Japan and around the world. NTT has very strict controls and high quality standards; a company that can sell a telecommunications system to NTT can sell it anywhere in the world.

A watchful American management consulting company, Dataquest, argues that NTT has not really been privatized, it has merely become "quasi-public" because the Japanese government still owns 33 percent of it. In 1985–1986 NTT entered into a large number of joint research and development projects with American firms. In doing so, they developed a truly aggressive form of cooperation with long-term ambitious goals. The Japanese believe that acquiring basic knowledge from the United States, especially from Bell Laboratories and from IBM, is no longer sufficient for their purposes; they believe that research must originate in Japan under NTT's direction. Especially significant is the effort to link NTT's network with IBM's system network architecture. Japanese competitors worry that the alliance will easily dominate markets everywhere because it will set international technical standards. But such alliances between American and Japanese partners are inevitable.

The Japanese Style of Business

Japan's economy puzzles even the experts: what belongs to whom? Some family-style, almost old-fashioned business practices still prevail; an example is the *nakodo*, a professional industrial matchmaker who arranges marriages of industrial companies. Verbal agreements and joint financing abound. A few large, but very flexible, industrial groups seem to call the tune with their banks, business agents, and political influence. And yet they define areas of responsibility down to the minutest detail.

Peculiar opposites characterize Japanese industry. Chauffeurs drive executives to the big laboratories, while small-scale craftsmen walk to little shops in the alleys and back streets of the suburbs. There the workers labor day and night, sorting or assembling components. The incredible number of cottage workshops astonishes the visitor from the West. Carrying baskets on their backs, riding bicycles, or using small station wagons, these craftsmen bring their products to collection points. Many of them are totally dependent on the big concerns, which pay them a pittance.

How have the Japanese succeeded to the point where they can pump 20 percent and more of their profits, year after year, back into research, development, and new production means? In 1975 America suffered one of its typical, frantic economic setbacks. Investments were down, interest rates were oppressively high, and the stock market slumped. As a result, many Silicon Valley employees were let go, and many firms decided against new acquisitions. The Japanese say that this was America's fatal error. During those years Japan had continued to expand, and when the semiconductor market took off again, they were able to deliver. In 1982 the six most important Japanese firms, wanting to establish their market share with the new metal-oxide-silicon chips, invested more than 20 percent of their sales in new facilities and equipment. This was an even higher investment than that of 1978, already an incredible 14 percent. American firms, in contrast, invested much less. The succeeding generation of silicon chips, which can store 256 thousand bits of information on the smallest crystal surface, could cost each firm up to 200 million dollars. Where does one find so much money?

Japanese firms have less government red tape to deal with than their counterparts in the United States, but business dealings are complicated by rituals and barriers, examinations and checks. Some years ago a former doctoral student of mine who was working in a large Japanese electronics lab, wrote me in desperation: "Here I need days to conduct an experiment that is a bit out of the ordinary. I wanted to leave a furnace for my crystals on all night. I had to collect fifteen stamps, which are the signatures here, from all different kinds of people. No, the Japanese certainly won't make it like this." Today, though, he has a different opinion. Despite the complicated social etiquette—perhaps even because of it—the divisions in the semiconductor war are resolutely marching toward the same clear goals.

Japanese unions, though tame and decentralized, still produce many strikes and contribute to the loss of considerable man-hours for Japanese industry. Each company has its own union, but the employees' close ties to the company, as well as their job security—at least in the large firms—help prevent large confrontations. Strangely enough, some strikes seem to come at the most favorable time for the employer, just when business is bad and warehouses full—times that in other nations would mean reduced hours. The payment system creates significant incentives for workers to cooperate with company policy. Only part of an employee's pay is a fixed salary; the rest is in premiums that are adapted to the firm's current economic state.

At international conferences, the Japanese researchers get together at night to exchange information and plan strategies. They gather much information from patents, which they pounce on. A properly written patent must first explain the present state of the art and the weaknesses of existing methods, then describe the new invention and its procedures. Japanese industrial employees are also supposed to generate their own patents. In many laboratories a young researcher can hope to attend a conference or travel to the United States only if he has written his two annual patents. Applications flood into the overtaxed Japanese patent office. People like to compare the number of patent applications as a measure of a nation's technological strength. But mere statistics do not reveal the level of quality of the patents. Nevertheless, using patents as a measuring stick continually reminds the researcher to formulate his thoughts clearly, familiarize himself with the literature, and consider what should be improved.

Patents, of course, do protect new techniques. To use devices and methods protected by patent, a firm must buy a license. Electronics and microelectronics contribute by far the largest number of patents. In 1980, 92,000 were registered throughout the world in these fields, compared with only 15,000 registered for chemistry, the runner-up! The international balance of trade for patent licenses is negative for Japan and for West Germany, both of which pay fees to use others' ideas in making their products. But as in all other branches of modern industry, the pendulum for semiconductor technology patents is starting to swing back toward Japan.

Creativity at Any Price

In Japan, industry recruitment of "brains" from the universities is a serious undertaking. In the summer of 1984 some of my scientific colleagues in Tokyo, as well as industrial managers and representatives of MITI and Monbusho—the Ministry of Education—informed me of their long-term plans. Roughly half of all graduate students are concentrating in the engineering sciences, and an additional 10 percent in the natural sciences. Almost two-thirds of the master's degree students serve as a rich resource for the state- and industry-planned expansion of technological fields. Costs of education are high, so the number of faculty members can be increased only if the number of students in that field increases. By arrangement with MITI and industry, subjects pertaining to technology are allowed additional staff.

Government research institutes are held tightly in check. At present, the state employs about 40,000 researchers, and this number is not expected to increase before the turn of the century, although there may be some regrouping in newer fields, with special attention focused on growth industries. The number of industrial researchers and development engineers will double by the end of the century—from the present 200,000 to about 400,000, and this makes American competitors panic. These plans are backed by a strategy that is very different from that of Europe, where government institutes have great difficulties in transferring technology to industry.

The result of this strategy of strengthening Japanese industrial research at all levels is immediately apparent. When I arrived at the Narita airport to take part in a scientific conference in 1985, I was met by a small group of friendly colleagues, whom I had never met before. They asked me to add a visit to their new labs to my schedule. A few years ago their plant was an old-fashioned glass factory supplying low-tech bottles and window panes. But now the laboratory, furnished with new and expensive equipment, was a researcher's delight. They aimed to go into optical communications with new materials and novel techniques. What did I think of it? they wanted to know. Would I be willing to cooperate on basic research? How about sending a young physicist or two to our labs in Germany?

Japan's innovative research has shifted from state-supported laboratories to industry. At least seventy new industrial basic research laboratories were to be installed by 1988, with the total investment exceeding two billion dollars. A long list of innovative projects is under way, mostly in materials sciences, including ceramics and artificial diamonds.

In the laboratories of Japanese universities one hears cautious criticism and subtle skepticism about this shift in emphasis. The priority given to industry as the site for basic future research reveals a weakness in the Japanese system right where new discoveries must be made. The rigid discipline of the industrial lab leaves little room for new ideas, especially during times of economic duress. Will the Japanese perhaps make the mistake of reining in the universities and paving over the academic playground?

Japanese politicians already know that not just the silicon crystal but other new materials must be mastered. A top-ranking politician of the governing Liberal Democrats, surrounded by an entourage of industry managers, recently visited European research sites, including the Stuttgart laboratories of the Max Planck Institute. We were impressed that this politician saw his political future ensured by keeping an active interest in materials sciences. How different from American and European politicians, who carefully and almost proudly try not to acquire the appearance of a scientific specialist or even an amateur!

The Americans accuse Japanese industry of not being creative, and this is true, or at least was true, as the Japanese themselves realize. No important scientific discoveries concerning crystals and electrons, except perhaps the tunnel diode, came from Japan. Investigating highly doped germanium, Leo Esaki found that electrons do not have to jump over a threshold constructed in the crystal but can travel through it as through a tunnel. Particles also traverse the nucleus in radioactive decay, a phenomenon that only quantum theory can explain. Esaki received the Nobel Prize for this research and today works for IBM in the United States.

When Japanese delegations visit the Max Planck Institute, they have one paramount question: how can we organize creative research? The Japanese are becoming increasingly aware of this need, which goes hand in hand with their success. Individual creative talents need to be cultivated if Japan wants to reach the top and maintain that position. The country's geographical insularity and its emphasis on the common good fuel attitudes that can hinder unusual ideas and courageous—even stubborn—individuality.

The Japanese passionately embrace everything concerning education. The educational system is based on highly selective tests and a rigid pecking order. There are plenty of personal computers in the schools, and contests are held all over the country for research and amateur science projects. Young people are taught about electronics systematically; a career in technology is highly regarded. At Japanese colleges and universities, technical and scientific subjects enjoy great prestige. And no other technological area so strictly demands and highly rewards the virtues of stamina and patience, care and cleanliness, discipline and precision as does semiconductor research.

NHK, the public radio network, broadcasts outstanding science programs. And Makoto Kikuchi, the research head of Sony, is well known to young people because of his appearances as a physicist on many television shows. In a book about Japanese microelectronics, Kikuchi writes that the Japanese, because of their upbringing, education, and dedication, are better qualified than Americans to make the best use of semiconductor crystals. "It was an incredible achievement to go from the 99.9 percent pure crystals that we had after the War to the necessary 99.99999999 percent purity—and in just one step. It was more than just cleaning the material and mastering the required chemical processes." It was a fanatic devotion to precision.

Japan's young people are very enthusiastic about all the new electronic games and devices, which makes for a strong and receptive domestic market. The microelectronics market in Tokyo's Akihabara district has become a major attraction for Western tourists. Foreign visitors are always surprised to see how expensive Japanese products are in stores on the Ginza. When they ask whether the prices of Japanese goods sold abroad are artificially low—"dumping" prices—the answers are usually complicated, involving direct and indirect sales taxes. In spite of the high prices, young Japanese pounce—albeit critically—on all the new products.

The Integrated Society

Makoto Kikuchi still lives in the same tiny house he had while a young physicist at the Electrotechnical Laboratory. Now, as director of research at a large corporation, he could easily move into a larger house, but that would tear him out of his community, and Kikuchi prefers to remain the way he is. Japan's society is more homogeneous than Americans, or even Europeans, could ever imagine, and this homogeneity causes all of Japan's competitors to worry. One of my colleagues once worked in Japan as a guest researcher in the lab of a firm that employed almost 100,000 people. He was given a list of all the foreigners working there; he was the sixteenth. The other fifteen were all English teachers, hired because of their correct accents. There are almost no foreign workers in Nippon, nor are they actively sought. Only recently has this xenophobic attitude begun to change as world markets call for a more cosmopolitan view. Now all the large electronics companies offer postdoctoral positions to foreign scientists and send their young staff members abroad. Industrial foundations donate money to foreign research organizations to signal that Japan is a member of the international research establishment, and a very rich member at that.

The concept of an integrated society dominates all the activities of MITI and other governmental bodies. Management in business and industry also continually promotes integration, cooperation, and mutual security. Workers and their bosses devote a lot of time to discussing new methods of operation. Everyone is informed and asked for an opinion right from the start. Yet anyone who has worked in one of these labs knows that these conventions are often merely rituals. The decisions have already been made, but the mutual discussions serve to unite opinions and smooth ruffled feathers. Most Japanese view "their" firm as "home" and even call it home.

The evening of my visit to a Japanese laboratory I met a young German physicist, a former student of mine who was working as a guest in a Japanese industrial lab. He showed me a pocket calendar the firm had distributed to its employees, which highlighted the first Wednesday of each month with a green box. On these Wednesdays, the bosses had said, everyone was supposed to go home at quitting time. The workers had become so dedicated to their work that they continued to pore over experiments, read the literature, and engage each other in heated discussions long past quitting time. No one wanted to be the first to give up and go home. Employees also vacation together in special hotels where they can enjoy their hot baths, with all expenses paid for by the company.

Sociologists believe that the traditional cooperation among Asian rice farmers, who had to work together on vast irrigation projects, is continuing in industry. Mutual respect and consideration of the country's welfare have not lost their significance. The Japanese gains recognition through dedication to joint tasks. In the research labs, it is not the highly intelligent lone researcher who makes the discoveries but the troops of dedicated employees.

Although Japan has been successful in many economic areas, every analysis of Japanese business puts microelectronics in first place. To capitalize on that position, the Japanese contracted with a Madison Avenue advertising agency to spread the word on their superior technology in the most important American magazines and journals. The cover picture of one of these full-color supplements featured a Kabuki actor in traditional white makeup, symbolizing Japanese tradition. Balanced on his index finger was the new symbol: a silicon chip, a little piece of crystal, the emblem of excellence in technology.

Yet modern electronics will change and endanger Japan's old cultural values. The word processor, called *wapuro* in "Japlish," is just one example. Clever new computers accept the simple spelling mode of the modern phonetic kana syllables. The traditional Chinese kanji characters are much more complicated and harder to memorize. The wapuro will show all kanji characters of the same spelling, from the most frequent to the less commonly used ones. This simple electronic help devalues the old virtue of memorizing kanji characters. It also makes it much easier to read the word processor screen. Laziness finds microelectronic collusion, just as it does in the supermarket, where computerized cash registers have rendered manual calculation—and even stock taking—obsolete.

9
Whither America?

AMERICANS feel they have put up with quite a bit from the Japanese. First Japan ruined the American steel industry, and then the textile industry also suffered. Even more of a blow were the swarms of little Japanese cars that journeyed across the Pacific, decimating Detroit and other centers of the automotive industry. Soon Japanese televisions and stereos boasting American-sounding names, such as Pioneer and Panasonic, were dominating the American market. But Americans kept their faith in free trade. The United States has a long tradition of open markets, and the government is reluctant to burden consumers with import taxes to give an advantage to domestic companies. An American politician risks losing votes if he calls for strict tariff barriers or federal assistance for business.

Bitter Battles

In 1981 Hewlett-Packard, one of the first Silicon Valley companies, needed a large supply of high-quality memory chips for its measuring instruments and computers. When the company checked the quality of the chips available, American suppliers made quite a poor showing; the Japanese chips were far and away the best. The telephone company, which needs reliable connections and exchanges that will operate around the clock for many years, made a similar survey. The military also took a look at the palette of offerings for its modern weapons. The results were embarrassing: only the Japanese chips measured up. America felt outdone and ashamed.

Having cursed General Motors, driven little Toyotas, and played

Sony cassette players, Americans had reached the limits of their tolerance for Japanese competition. The transistor was a great American invention, and semiconductors were an American domain. Silicon Valley was the last Western frontier of the American pioneer spirit.

Americans assumed that the Japanese had merely exploited the market through rock-bottom prices, unfair dumping practices, and sneaky export offensives rather than technological know-how. The argument that Japan offered higher quality and not simply lower prices was considered untrue. But comparisons proved that the Japanese chips had fewer malfunctions, lasted longer, and displayed more suitable characteristics. Most important, deviations from one sample to the next were much smaller. Japan had a much better understanding of the crystal.

American engineers felt that the semiconductor war was being conducted unfairly. Organizers of international conferences on integrated circuits were repeatedly embittered by the lack of response from Japanese researchers who were invited to deliver papers. The Japanese, it appeared, just wanted to listen, not report on their own work. Only in the 1980s did they make an effort to deliver papers and organize bilateral conferences. Japan's craving for information did lead it to employ some questionable methods. Silicon Valley was shocked when agents working for Hitachi were caught trying to bribe their way to American high-tech secrets. Eighteen high-ranking Japanese businessmen were convicted of having arranged to have secret documents stolen from IBM. Two Japanese computer firms, Hitachi and Mitsubishi, had reached an impasse in their competition with IBM. At the time the IBM 3081 series posed an almost insurmountable obstacle to Japanese expertise. While they had few problems with the hardware or with silicon, they could not master the system's architecture and software. The two firms thus concocted a plot to steal information by using shady wire-pullers and fictitious firms. Originally chasing Soviet spies, the FBI caught six Japanese, including one of Hitachi's chief engineers, Kenji Hayashi. In February 1983 the six admitted their guilt in court. The sensational case was quickly given the epithet "Japscam," the first syllable of which served as a pejorative reminder of Japan's role in World War II. Up until then Americans had politely avoided using the old term "Jap"; the conflict had erupted anew.

Indictments and counterindictments traveled from one high court to the next. A new round of trials involves two engineers, Raymond Cadet, previously with IBM, and Barry Saffaie. In Mountain View, right in the heart of Silicon Valley, these two suspected burglars were apprehended by the FBI with the help of video recorders. Electronic espionage and counterespionage!

Reports from the semiconductor battlefronts—the laboratories, the dustproof wafer fabs of the silicon foundries, and the courts—are big news in the United States. In the spring of 1984, when Hewlett-Packard reported on the next generation of silicon memory chips, dozens of reporters were present. Their new hp1000 minicomputer required huge numbers of the newest 256 thousand-bit memory chips, which had to be of the highest quality and reliability. Bob Frankenberg broke the good news first: the American chips were three times better than they had been four years earlier. A sigh of relief was heard in the auditorium. But then for the bad news: the Japanese had improved sevenfold. To supply the memory devices, Hewlett-Packard again turned to three Japanese firms.

A lot of work lies ahead for American labs and manufacturers, who cannot afford to compromise on the slogan, "Zero Defects." The crystal must be treated with such precision that absolutely no defect can arise in its structure. And the Japanese have a lead of at least one year's time in the struggle to perfect silicon.

In the eight years between 1977 and 1984, the United States lost no less than 20 percent of the worldwide market for integrated circuits to the Japanese. The mass market for memory chips has virtually become a domain of the Far East. With each new generation of these most important computer components, the Japanese have secured their rule. Charlie Sporck, head of the National Semiconductor Corporation of Santa Clara, California, conceded in a speech that "the dynamic random-access-memory market . . . is now a virtual Japanese monopoly."

The Government Gets Involved

The armed forces in particular worried about the decline in American technology. Complicated weapons systems, with increasing numbers of electronic components and increasingly lethal potency, must be as foolproof as possible. They require ever greater automation in operation, in checking, and in monitoring. Many are built to meet the "fire-and-forget" requirement. Demands for reaction speed are becoming tougher all the time. Observation, analysis, and information exchange require split-second timing. The Pentagon calls for rapid, infallible circuits that will hold up on dirty battlefields.

To fulfill its military needs, should America import the best silicon chips from Japan? The Pentagon is in a real bind. Fighting "wars" on two fronts, the United States can't trounce the Soviets if it capitulates to Japan. In the spring of 1984 Richard de Lauer, an undersecretary in the Department of Defense, made a surprising statement at an electronics engineers' conference on a "national electronics strategy." "The Russians," he said, "are incorporating our American technology in their new weapons systems faster and better than we are." He went on to explain. "Everything takes too long here. We need an incredible amount of time for red tape and all the decision-making processes. The endless checking and price determination of new components are the major barriers keeping new technology from rapidly becoming a part of the American weapons system." He described how the Russians stop at nothing to obtain silicon products made in the United States. Making just a few small changes, the Soviets can incorporate these chips in their own armaments without months of pretesting. In response the Pentagon has considered undertaking its own research programs to stop the leaks of both information and integrated circuits to the Soviet Union. The move may prove unpopular because it will cost a lot of money and render governmental agencies very powerful in an industry that has been known for its fast pace, free markets, and open international trade policies.

The Pentagon needs very high speed integrated circuits (VHSIC). This acronym, pronounced vee-sick, has become a catchword. American companies have been challenged to meet the stiff requirements of the Pentagon's projects. Right from the start America informed its NATO partners that everything concerning VHSIC had to stay at home in the United States! The smaller firms in Silicon Valley participate in these plans hesitantly, still putting their faith in market forces. The Pentagon's most important partners are the airplane and spacecraft concerns, for which the military market is vital. These businesses were the first to agree to the VHSIC project and step up their silicon research.

Other battlefronts pose problems as well. Very large computers are in great demand by the military for analyzing weather conditions, enemy codes, and ballistics trajectories. In addition to current weapons systems, the Strategic Defense Initiative, or "Star Wars," will rely heavily on huge computers and fast semiconductor hardware. The military significance of the massive and controversial SDI program precludes the use of imported chips, which exacerbates the crisis posed by the silicon crystal. Despite their crucial role in national defense, the biggest computers are not made in the United States. One that is available, named for Gene Amdahl, a former engineer at IBM, is manufactured primarily by Fujitsu in Japan. Recognizing the need for American-manufactured large computers, in 1981 the Pentagon presented to the U.S. Congress its plans to build them. Heated debate ensued. Upholding individual initiative and private business, members of Congress ridiculed the Pentagon's wish for its own large computers, terming it a "road show." At present, however, it appears that the Pentagon will pour hundreds of millions of dollars into semiconductor research and technology.

Several government controls in the United States keep East Asian competitors from gaining a competitive edge and costing Americans their jobs. The Western nations have established a coordinating committee, CoCom, which aims to stop the illegal export of strategically important goods, especially from the United States, to Warsaw Pact nations. In the spring of 1984 strict controls and regulations were established for the first time. There are many channels through which semiconductor elements, software, or production instruments for silicon technology can fall into the wrong hands. Small circuits can be carried in diplomats' briefcases. Larger items are sent via Helsinki, Vienna, Belgrade, and other cities. As a result, however, the ones who suffer most from the new controls are the Western allies, who feel hindered by growing red tape and left out of the newest technology.

Appealing to Uncle Sam for aid is a painful last resort for an American businessman. But desperate cries for help have recently resounded from Silicon Valley. Charlie Sporck, president and chief executive officer of National Semiconductor Corporation, the Valley's largest silicon company, delivered a dramatic, alarming address in San Francisco in November 1985 before WESCON (Western Electronics Show and Conference), the industry's important annual meeting. One of his employees, who had experienced the decline of the American radio industry in the 1950s, told him that the same fate seemed imminent for semiconductors. Sporck warned, "Complacency means certain death." A few years back, no one would have expected such a pessimistic speech from this self-confident industrial pioneer. Even less would one have expected his cry for government protection of the microelectronics industry.

"Let's face it—protectionism beats extinction any day," he said, imploring the government to install trade quotas and levy stiff import duties on all Japanese electronically programmable memories. He also called for better conditions for capital and improved relations between management and labor.

But Sporck's pleas and those of others brought little action in the nation's capital. In a meeting in the spring of 1986, U.S. trade representative Clayton Yeutter told the industry people that he did not condone accusing Japan of unfair trade practices in electronics. Although he used strong words describing the Japanese "intransigence," he opposed any changes in Regulation 301 of the Trade Section, which gives the administration broad authority to institute countermeasures against unfair trade practices of another nation. Instead of immediate sanctions, government officials proposed long-term solutions, the use of international agencies, and redress under the General Agreement on Tariffs and Trade for the endangered semiconductor industry. But other developments eased the situation temporarily. In late 1985 and again in 1986, the yen began to climb sharply against the dollar and other currencies, making Japanese exports costlier. Semiconductor producers in California, Arizona, and Texas enjoyed a reprieve.

The first of the five legislative priorities set forth by the U.S. Semiconductor Association is a tax break for research. The second is fair regulation of exports, keeping national interests in mind without strangling trade. The third priority is regulation of tariffs and duties that will not burden the reimportation of chips finished in the Far East. The fourth is a ruling to except research mergers from punitive antitrust measures. The fifth and final priority is the establishment of legal protection for software programs that are vulnerable to theft. This list of controversial proposals shows how the new industry has gradually awakened to reality. A new joint effort, headed by Bob Noyce, has been created: SEMATECH, a government-supported model production facility with the latest and best equipment for fabricating circuits of very high integration. All American semiconductor companies, it is hoped, will benefit from its pioneering efforts.

Flaws in Education

Surely one of America's greatest weaknesses in the race with Japan lies in its system of education. Most grade schools and high schools have weak science programs; some school districts do not even have a properly educated physics teacher. Of the approximately 200,000 employed mathematics and natural science teachers, half either have not studied these fields adequately or are unqualified to teach them. In the last ten years the number of mathematics teachers has dropped by 77 percent. No wonder, because a math instructor earns much less than a halfway successful insurance salesperson. Any teacher can earn more than twice his or her salary by working with software or computers in industry. American schools should extend their scanty math requirements by a year, but they would not even be able to drum up the minimum 68,000 math teachers to do so.

The situation at the universities is even sadder. Students who miss opportunities or motivation in grade school or high school will very likely decide not to tackle difficult majors such as math, physics, or computer science. An MBA graduate can earn much more money than a young technologist anyway. Why in the world would anyone want to spend years studying one of the sciences?

If American students are rushing into business classes, who will work in the excellently equipped labs of the elite universities? Significantly, almost half of the doctoral students in the engineering sciences are foreigners, mostly Asians. Young Chinese from Taiwan, Indian software specialists, immigrants from the Philippines, and refugees from Iran, as well as many Europeans, are tackling engineering. They know their skills will be needed. That is the key to getting a "green card," which allows them to work. Many of these young students may return home, where they will be rewarded with social prestige and economic success. America, then, faces a brain drain of its own—the loss of well-trained minds to Far East competitors.

The total number of young specialists in the United States is already dangerously low: of 10,000 college graduates, only a scant 70 are engineers, compared with 400 in Japan. In the 1980s the number of students receiving doctorates in the technical fields was a third less than ten years before, and it continues to fall. The situation is especially precarious in the computer sciences. Schools simply cannot recruit the instructors or professors they need. Private universities offer prospective teachers special grants and aid, loans for homes, and consulting contracts with local industry. The best teachers make much more money in an industry begging for their services.

The lack of specialized personnel will set limits on American technology, but the scales may swing the other way in the future. Technicians' salaries will rise in relation to those of professionals in the increasingly overcrowded fields of law and business. In addition, schools can expect to mine talent from a new generation of computer enthusiasts who have learned the technology through home computers. This will all take time, though. In the coming years, America will be tempted to import experts from all corners of the world, but its immigration laws will prove a stumbling block. Only heavy pressure and complicated red tape enable exceptions for scientific experts.

The universities are fighting to maintain their standards by seeking government financing for new, expensive equipment and by offering totally new fields of study. Some universities are forging new links with industry. North Carolina, for instance, constructed a research park in the triangle of land bordered by its three major universities: Duke, the University of North Carolina, and North Carolina State University. The park is the headquarters for the Semiconductor Research Corporation, sponsored by such companies as General Electric, IBM, AT&T, and Dupont. Microstructure sciences, semiconductor design, and device manufacturing are taking the lead in the joint effort to support academic research and training. The three universities are cooperating to expand their science and computer programs. In addition to their university contract, professors receive a guarantee of compensated extracurricular work. The University of California at Berkeley has also started a large materials research center that will be used by several faculties and will assist many industrial firms.

Cornell University used a major grant from the federal government to establish a national center to study the smallest structures. Researchers will work with new sources of light to engrave fine circuit designs onto semiconductor wafers. A race is under way to invent special tools for transferring the minute designs. With circuits packed so closely together, however, unexpected difficulties, as well as possibilities, may result. Smaller than a micrometer, which is only a thousandth of a millimeter, these transistors will also be researched at the Submicron Facility, the lab center for scientific measurements.

I once spoke at the opening of a microelectronics conference in Aachen, West Germany. The Cornell people pricked up their ears when, in praising microstructuring, I quoted Gotthold Ephraim Lessing's epigram: "Was artig ist, ist klein," meaning, loosely, "good things come in small packages." The poetic German phrase was later inscribed with an electron beam; the width of the letters was only a few atomic distances!

Other universities are eager to follow Cornell's example. Using funds from foundations the state of Texas is establishing laboratories and professorships, with the aim of retaining not only the people trained there but also the semiconductor industry. The Ivy League universities are also entering the competition.

New Avenues for R and D

Cooperation between competing industrial firms has strengthened remarkably, and some novel approaches have been taken. Twelve companies founded the Microelectronics and Computer Technology Corporation in Austin, Texas, to meet the technological and economic challenges of the Semiconductor Age. The group now comprises more than twenty companies, including some very large ones. Each member has the right to use the resources in any one of the cooperative's programs and to send people to Austin to do research. Members share the burden of financing expensive development—a truly remarkable deviation from the traditional American spirit of going it alone.

By 1987, however, many companies were so hard hit that they had to withdraw from the cooperative ventures. The future of the laboratories was seriously endangered. Making the situation even more touchy, Japanese companies applied for membership. American members debated giving them a cold shoulder even though they ran production lines within the United States.

But what about the small and medium-sized firms that manufacture semiconductors, software, instruments, and devices? Not every manufacturer can afford to do its own increasingly costly and complicated transistor production. For this reason the universities have been able to attract smaller firms to join research consortiums. Or such firms have founded clubs, in which fees cover research and development, and results are made available to all club members. The cooperative approach has allowed them to keep up with the latest technological developments.

The antitrust laws do pose a threat to these efforts, but experts hope that mergers in research and development will continue to be permitted. Firms working with the silicon crystal are seeking preferential legal treatment, which the courts will most likely grant. The government will probably allow other liberal merger incentives for the silicon industry.

Smaller firms deserve government support in their efforts to finance new research and development, for such firms have scored some victories in the semiconductor war against Japan. Just when everyone believed that inexpensive Japanese pocket calculators had taken over the entire world market, William Hewlett and David Packard decided to construct a much more sophisticated pocket calculator equipped with logarithms and exponential functions and the complete array of complicated statistical arithmetic operations. When engineers began buying this silicon slide rule, the high-performance calculator's success was assured.

The all-important concept of the microprocessor was the brainchild of Apple, a small company that was just starting up. The Apple Company became a symbol of the hope America places in its young people. Its founders, Steven Jobs and Stephen Wozniak, wanted to use sophisticated silicon chips to make their own computer. Why, they argued, shouldn't everybody have his or her very own good, small computer? The machine they envisioned was for everyday use, no longer a giant that had to be run by experts in climate-controlled, specialized rooms. To the surprise of everyone, from Japanese marketing experts to big computer moguls at home, the two men succeeded in developing a home computer in a garage right in the middle of Silicon Valley. With a good dose of courage, the Apple Computer Company took the lead in the industry with a personal computer that was technically problem-free and easily marketable. Public demand was so great that Apple could scarcely keep up. The big firms soon jumped on the bandwagon of personal computers. Congress even passed a tax rebate to enable schools to buy computers at lower prices, greatly expanding the company's markets. Other American and Japanese companies, small and large alike, soon followed in Apple's footsteps, but it was the two young Californians who discovered the new market for personal computers.

Wozniak and Jobs became millionaires and symbols of high-tech success. The computer had finally shed its aura of inapproachability and of government centralization and control. This change in attitude was important, for in the early days, computers were still thought of in George Orwell's Big Brother imagery, as tools of the ruling bureaucracy, the military, or other centralized government bodies. In factories and businesses, the computer was the omniscient central brain to which only a select few had access. You had to master a special language just to communicate with the machine by means of the visual display unit. With the advent of personal computers, this threatening image was destroyed. The democratization of the computer, made possible by cheaper silicon memory chips and microprocessors, continues unabated.

The End of a Monopoly

The East Asian challenge activated not only the government and the nervous little firms in Silicon Valley, but also the powerful East Coast giants. On Old Orchard Road in Armonk, New York, IBM began formulating new policies, and on Broadway in Manhattan, AT&T started making strategic plans. These two powerhouses were threatened by each other as much as by Japan. Because of the great advances in silicon technology, computers and telephone systems are now so similar that the firms' previously peaceful coexistence has turned into tough rivalry.

For many years, though, both companies were jeopardized by strict antitrust legislation. After investigating trainloads of documents, the government dropped its antitrust case against IBM. The conclusion that IBM did not have a clear monopoly was supported with evidence that small firms were doing very well and that newcomers enjoyed a good chance of success in the computer market.

But the telephone company case dragged on. With more than a million employees, a record number of stockholders, and annual sales of $69 billion, AT&T was one of the largest industrial concerns ever to be summoned to court. Up until then the government-regulated monopoly had been required to provide telephone service all over the country, limit its operations to telephone business, and release information on technological developments such as the transistor. But in 1978 a number of smaller firms accused AT&T of keeping everyone else out of the lucrative phone business, arguing that they could carry out some of the communications tasks much more cheaply. Bell, they added, had given its affiliate Western Electric a monopoly on telephone equipment, cables, and telephone links. No other supplier stood a chance of getting even a crumb of the giant cake. AT&T's technical standards alone gave it an unfair edge.

Most Americans enjoyed Bell Telephone's good service and reputation for quality. Foreigners repeatedly confirmed that the American telephone system was much cheaper and more efficient than the service in their own countries. And scientists pointed out that only a company the size of AT&T could conduct the high-quality research that had led to the development of the transistor, various Nobel Prizes, the first communications satellite, the solar battery, and many other advances. Technologists and engineers worried about the effect of the case on future progress.

In 1983, Judge Harold H. Greene announced that AT&T would be broken up at the end of that year; only a skeleton firm with that name would remain. Western Electric, the production company, and Bell Laboratories would continue as part of this organization, but Western Electric had to give up its equipment monopoly. Seven regional holding companies were created to carry on AT&T's telephone business. The telephone market was thrown wide open.

The regional Bell operating companies realized from the start that only with first-rate research could they stay in front. They formed a new organization, Bellcore, to research and develop new devices and systems, aid local companies in choosing equipment, and help plan the networks. Many scientists and engineers from the old Bell Labs accepted jobs in the new organization. The original laboratory in Murray Hill, New Jersey, the cradle of the first real silicon crystals and transistors, underwent a long, gloomy period of adjustment to the breakup of its domain, a wrenching process for everyone who had experienced that unique research atmosphere. On New Year's Day, 1984, the two labs, AT&T's Bell Labs and Bell Communications Research, were officially set up to do telecommunications research. There will undoubtedly be competition and overlap of work instead of cooperative division of labor under one roof. Pressure from the factories will undoubtedly endanger both the freedom and the drive to do long-range basic research.

Experts sharply criticized the breakup of AT&T. *Time* magazine summarized public opinion, stating that the breakup contributed "to the decline of Western Civilization" (Feb. 2, 1987). In Silicon Valley, Gordon Moore, a respected spokesman, termed the situation a national tragedy. Breaking up, he said, destroyed a national treasure. He emphasized that Bell had always generously shared information with all the newer firms, which then grew and thrived. Who would now finance a team dedicated to new frontiers, risky ventures, large-scale research? Moore complained that California would have to look to Musashino, where the central labs of the Japanese telephone concern NTT were expanding. Only recently have many civil servants and politicians become worried about the severe pruning of Bell Telephone's potent research. Experienced congressmen have admitted to me that this may have been a serious, strategic mistake that could weaken the United States, but the overwhelming majority in the Senate and the House of Representatives was on the antitrust side.

The Japanese realized that the breakup of AT&T changed the prospects of the international communications industry. In just a few years, they predicted, this sector would outdo the energy industry. Most experts agree that communications is the business of the future.

The ruling also allowed AT&T to shed a straitjacket. The company no longer had to refrain from marketing hearing aids, calculators, and other nontelephone equipment. The new AT&T was free to enter the world markets. The market was unexpectedly enlivened, and great opportunities arose for every bidder.

AT&T's accumulated expertise could blanket Europe and other markets, although the company still lacks a strategically organized sales network, never having needed one before. In the Netherlands it signed a pact with Philips to manufacture and market telephone systems. To make and sell smaller equipment, AT&T, after a long search, chose Olivetti, promising a bright future for the Italian firm. Spain was chosen as the site for at least one very large silicon factory to produce AT&T's latest integrated circuits. These moves prove that the firm has moved out of its American confines. With its estimated three billion dollars' worth of annual semiconductor production, AT&T has sharpened its rivalry with IBM, which boasts estimated sales of one and a third billion dollars. The two companies have extended their competition to European and Asian markets.

AT&T's Allentown, Pennsylvania, laboratory, long treated as a stepchild, has evolved into a major center of silicon technology. There, two hours by car from headquarters in Murray Hill, the vice president of electronic technology, Klaus Bowers, born in Germany and educated at Oxford, established "Bowers' Towers" for semiconductors. His team developed memory devices intended for the open market rather than for use just by the telephone company. But selling them proved difficult because of Japan's market dominance.

Beginning in the eighties, a new MOS memory technique was put into practice. The technique combines two kinds of transistors: those in which the electrons snake through a controlled channel and those in which electron holes traverse it. It is called CMOS, complementary metal-oxide-silicon technology, and it has beaten competing technologies. In this technology defects may arise at many stages, lowering the yield of functional circuits. But it is worth the trouble, for the skillful coupling of the two kinds of transistors saves energy. In the inactive state, during which almost no electricity flows, the circuits remain much cooler than their predecessors. Denser and smaller memory chips, however, require more thrift with energy, for if a semiconductor crystal becomes too warm, the atoms in its structure begin to move rapidly, breaking the bonds. Electrons then flood into all the finely constructed structures, erasing all the information stored in their memories. Researchers are working to perfect their control over the flow of electricity and its impediments.

New tools are being used to investigate silicon. Scientists are trying to find a way to repair lightly damaged memory chips to raise the yield of usable elements on the silicon wafer. Manufacturers now equip the wafers with a few extra memory cells right from the start, to cover loss. After traveling through the furnaces and machines, each element is precisely measured by a large computer, which uncovers every defect. The silicon chip in the quality control computer checks the next generation faster than the human eye can follow. In just fractions of a second, the computer decides how a defective memory chip can be repaired and commands a control device to send a strong laser beam to the point in question. The ray severs the metal bonds and alters the switch. The defective part of the circuit is then replaced by a flawless piece from the reserve supply.

After many years of Silicon Valley leading this field, many firms on the East Coast now feel they are back at the top. AT&T hopes its new breakthroughs will turn into profits on the international market. While its laser surgery impresses the trained observer, AT&T's closed system for mastering silicon is perhaps more important. Unix, as Bell calls its rapid, high-performance computer system, will improve the new digital telephone systems and set a new industry standard. But to penetrate the world market, the company will have to learn how to sell on open markets and will also have to reduce the number of employees by more than a quarter.

IBM, or Big Blue, has not sat back quietly during the last few years. Both its researchers and its sales teams have chalked up unexpected successes. When the two "schoolboys" at Apple introduced the home computer, people smirked at IBM's embarrassment. But Big Blue retaliated with its own personal computer, which soon put the cheeky newcomer in not-so-glorious second place. From a technical standpoint IBM's product was nothing novel, but the total package, with its wealth of software programs, IBM's good reputation, and guaranteed customer service, gave the company the edge. In almost no time these traits allowed IBM to lead the market in minicomputers as well. By the beginning of 1984, IBM was putting the finishing touches on one small home computer every sixteen seconds, and they could hardly keep up with demand.

In Fishkill, New York, and Burlington, Vermont, IBM now manufactures its own circuits for the logic components of its computers instead of buying chips from other suppliers as it used to do. The firm also produces chips at many other locations around the world. But competition is tough: "clones" of the successful IBM personal computers are made much more cheaply in the Far East, and American consumers cannot resist the temptation of bargain prices.

Light Communications

Of the wealth of research findings generated in the old Bell research labs, one of the most important was the use of light to transfer information via fine fibers. By the old method electrons traveled through copper wire to transmit speech signals from place to place or radio waves were modulated and sent from tower to tower. But light is a better carrier of such signals, for it oscillates faster than radio waves and can thus transport many more telephone conversations at a time. The disadvantage of using light is that fog, rain, or temperature fluctuations can scatter and weaken its signals beyond recognition. A light beam has to be guided and treated gently.

In solving this technological challenge, the silicon atom once again played a special role. The mineral quartz consists of silicon oxide, the compound formed by one silicon atom and two oxygen atoms, the two most common elements of our planet. According to theory, a quartz fiber thinner than a human hair can conduct flashes of light imprinted with information over long distances. The basic concept sounds simple, but it is difficult to realize. A glance out the window makes the problem clear. Even high-quality window glass, be it ever so thin, blocks quite a bit of daylight. To transfer light over many miles requires incredibly pure quartz glass. Imagine trying to peer through a pane many miles thick—you would not be able to see anything at the other end. Just a few water molecules in the glass accidentally would absorb a light beam and wreak havoc with the signal.

The incredible purity of silicon crystals that was gradually attained gave chemists and physicists the courage to attempt producing sufficiently pure quartz glass fibers. After determining the scientific laws governing the reducing and scattering of light in a fiber, they discovered that water and iron molecules had to be avoided at all costs. They also knew that the route of the light had to be set precisely. For that task, the oxide of germanium, the first and now long-forgotten semiconductor, played a key role. By admixing germanium oxide, experimenters produced the desired changes in light diffraction within the fiber.

Today a signal transmitted in a flash of light can travel through fibers more than 150 miles long without amplification before being decoded at the other end. Telecommunication cables have become thinner and thinner while maintaining a high transfer capacity. The rapid oscillation rate of light makes it suitable for transmission of dense amounts of information. Recognizing a big chance, AT&T researched light wave communications intensively for many years. It was given practical application in New York's Wall Street financial district, where the huge demand for phone hookups left no more room below the sidewalks for thick, conventional copper cables. Glass fiber cable proved to be the answer. Glass fibers are also used now in transatlantic cables. The cable sections are long enough that the number of intermediate amplifiers can be kept to a minimum. By early 1983 AT&T had installed a glass fiber cable the length of one light-second—186,000 miles, the distance light travels in a second.

Every technological advance has its cost. As a result of the change to glass fiber, the copper cable factories in Baltimore and elsewhere, with their many jobs in specialized metal processing, have closed. In their stead have arisen new factories working on the new technology. Once again, silicon triumphed over metal.

Protectionism or Cooperation?

April 1, 1984, was a historic date in the semiconductor war. On that day the customs barriers between Japan and the United States for semiconductor devices were eliminated. This bold measure surprised participants on both sides of the Pacific who had clamored for government protection from actual or imagined unfair competition. Adjusting to the newly opened markets, industries in the United States and Japan accelerated research and development, applications and markets. They will most likely be joined by the rest of the Pacific Rim nations—Singapore, Malaysia, and their neighbors—in tearing down protective trade barriers. Only one region of the world—outside of the sluggish Eastern Bloc—reacted differently. Taking refuge in its fortress, countries in the European Economic Community levied a hefty 17 percent duty on every imported electronic circuit, as demanded by European industries.

The open market strategy produced considerable legislative changes in Japan. Following the American example, Japan privatized telecommunications. Previously modeled on the European state-run telephone companies, the Nippon Telegraph and Telephone Corporation became private in 1984. New competitors entered the field. *Dai ni den-den,* the "number two tel-tel" company, went into operation in the spring of 1984. By chance, at that time I was visiting the labs of the once all-powerful NTT. Watching my hosts keep up the relaxed, confident appearance appropriate for a large research facility, I sensed that the new domestic competition would trigger faster, better research work.

Technology has become the hub of power politics in Washington and Tokyo. At the start of 1984 Trade Ambassador William Brock spoke before the House of Representatives' Subcommittee on Trade. With cautious optimism he reported on the progress of comprehensive negotiations between the United States and Japan. The United States pressed the Japanese to open their domestic telecommunications market to outside competition; NTT would be awarding contracts worth more than three billion dollars annually. Using massive pressure, the Americans succeeded in exporting 140 million dollars' worth of goods to Japan, but that amount represented only a third of Japan's exports to the United States. A three-year pact signed with the Japanese in 1985 provided for joint research projects. In addition American suppliers demanded a look at the research taking place in NTT's ever-expanding laboratories. The Americans will most likely get most of what they want and will thus be informed early on about technical developments in Japan in glass fibers, semiconductor components, and computer techniques. The two competitors have laid the foundations for working together to conquer the markets of the rest of the world.

Despite the cease-fire, the American trade war with Japan has intensified. The latest reports on rivalry in the telecommunications markets are unsettling to Americans. Even in 1983, when AT&T was still intact, imports of telephone equipment from Japan were growing. From 1977 to 1982 the United States easily maintained a satisfactory trade surplus in this important export branch, but in 1983 a deficit of 250 million dollars was racked up. The seven regional telephone companies carved out of Bell turned to Japanese products. In 1986 no less than 45 percent of all imports came from Japan, according to the Electronic Industries Association. Japan's share of the American market was almost twice as large as that of the runner-up supplier. The smaller American equipment suppliers were disappointed; they had hoped that the trust bust would give them access to the telephone market.

Dissolution of the monopoly made consumers unhappy also. When government subsidies were discontinued, the cost of local phone calls rose. Not only did people's phone bills increase but jobs in the industry were threatened. In November 1984, just in time for the presidential elections, the government proudly announced an unusually low inflation rate. But there was one obvious exception in the indicators: the cost of telephone calls easily led the index with an annual 16 percent rise.

The cut-throat race for the information market also put its stamp on domestic politics. The late Secretary of Commerce Malcolm Baldridge, a member of a Republican administration, had publicly speculated that the nation might need a unified industrial policy. His suggestion was not just a sign of a power struggle with the Pentagon, which now sharply monitors the trade of technological goods. It represented a change of attitude about government's involvement in industry. Impressed by Japan's united government-industry efforts and MITI's role, Americans are beginning to waive their free-market principles.

Louder and louder are the cries for harsh economic sanctions against Japan. In May 1986 the House of Representatives passed a stern bill directed mainly against Japanese imports. The president, however, saw this legislation as the possible beginning of a trade war in which all parties would eventually lose. Yet all peaceful attempts to assure self-restraint or voluntary import quotas for silicon devices have failed. American industry has accused Japan of dumping chips on the American market below production costs, a charge that is difficult to prove. No one can ascertain whether the Japanese are charging illegally low dumping prices. They maintain that their prices are low simply because of better scientific methods and cleaner factories. They also point out that Japanese firms are producing high-quality, inexpensive goods in the United States. In private conversations American experts admit that there is indeed better control, cleaner work, and higher yield in the Japanese semiconductor plants, and many experts say that in the future certain complicated integrated circuits may be producible only in Japan.

The promising results of Texas Instruments' memory chip factory at Miho, Japan, made many people in the semiconductor industry consider importing those chips from Japanese suppliers or starting their own production in Japan, the Silicon Island. Texas Instruments, which was the leader in semiconductor sales for many years before being surpassed by the Japanese NEC, is also credited with successfully maintaining a majority interest in a Japanese subsidiary. All other attempts by American firms to achieve control have failed, being merely joint ventures with minority ownership—another grievance for the American side.

In 1984, after much bargaining between the two countries, price levels were fixed high enough to make American silicon chips competitive. This system worked at first but fell victim to the "law of unintended consequences." Since chip prices in Taiwan and Korea remained low, and even dropped, American equipment makers found it advantageous to have their circuit boards mounted with chips in the Far East. As a result, jobs in the supply industries for silicon chips were lost to Asian firms.

If Japan is becoming increasingly efficient, how will the United States fare in the continuing semiconductor war? Wall Street's Paine Webber estimated that by 1988 the Japanese will have more than half of the world microelectronics market, which by then will amount to more than 50 billion dollars. What will happen to the United States' $11.4 billion semiconductor industry? The gravity of the crisis was apparent to Americans by 1986, when electronics had become the "biggest boss." Two and a half million jobs in this industry made it by far the most important employer of the nation, well ahead of transportation, food, and machine tools. Yet this industry was falling behind the Asian competition. More than 90 percent of the generation of memory chips with 256 kilobit storage capacity was already controlled by Japanese manufacturers. Generations of even denser packing, with 1, 4, and 16 megabits of information are being developed by several Japanese companies, and more and more American competitors are dropping out. The largest supplier of memory chips in the early eighties was Mostek, an American company, but within a few years it gave up making memory chips and was sold to a French electronics concern.

The Japanese have mastered a wide range of materials and components basic to the semiconductor industry. The ceramic platelets on which silicon chips are mounted for high-quality applications have become an almost complete monopoly of a company in Kyoto. The slices of highly perfected silicon crystals can now be purchased in Japan. The Japanese steel companies that have begun to diversify will soon be vendors of ultrapure silicon. Japan has also taken the lead in tools and equipment, which used to be an American specialty. High-vacuum electron guns to evaporate metals, laser sensors, and lenses of previously unheard-of sharpness, X-ray sources, and plasma machines define the new age of precision tools. The Japanese are producing equipment that is not just cheaper but technically superior.

A task force in early 1987 found that the United States had fallen behind Japan in seven of fourteen critical areas of semiconductor skills: processing, automation and production, packaging, materials and chemical supplies, and information management. The United States still held a clear lead in basic research in physics, chemistry, and metallurgy, in using computers for designing chips, and in the application of entire systems. The message is clear: The United States leads in invention and systems knowledge but has lost its ability to manufacture chips with the greatest efficiency. For this reason proposals are often announced for helping the semiconductor industry with some government-supported laboratory or corporation, especially in areas related to fabrication. One of the most active supporters of tax subsidies is National Semiconductor's Charlie Sporck; forced to abandon his own research department in the crisis year of 1985, he is understandably concerned about the Asian onslaught.

Silicon Valley observers pointed out that several American business practices formerly considered strong points were actually self-destructive. Quarterly financial statements, once regarded as a sign of quick adaptation and tough management, were seen as a disadvantage against the financially stronger Japanese and Korean firms. Unlike the American companies, those firms tolerated quarterly, even annual, losses in order to reach their long-term goals of microchip dominance. And the specialized independence of the Valley companies rendered them relatively weak against the large integrated concerns in Japan, and even in

Europe. The smaller firms could not afford to invest in large factories for the production of highly sophisticated chips. Silicon Valley had to admit its weaknesses.

The erosion of technical production expertise in America and Western Europe is most clearly apparent in microelectronics. America's stronghold, basic research, is no longer an automatic lead to technical superiority; it may even hinder it. Basic research tends to neglect practical problems and displace technical personnel from factories to ivory towers geared to the distant future. This counterproductive aspect of basic research is a bitter pill to swallow for Americans and Europeans alike. In response to the report of a Department of Defense task force, which pointed out significant weaknesses in American factories, the U.S. government has launched an effort to sponsor semiconductor manufacturing as well as research and development. The results of the government-aided SEMATECH program will be made available only to the participating American firms, certainly not to the Japanese competition and not to the European NATO allies. But what should be the status of Japanese-owned companies located in the United States? Such questions led to a painful awareness that the cherished American ideal of free enterprise and open competition, without governmental interference, has come to an end for the silicon chip enterprise.

The government hopes that semiconductor manufacturers will learn how to treat the silicon crystal under extremely clean conditions, fabricate fine patterns, and obtain high yields. The new techniques in manufacturing may have to rely on X rays to obtain even more intricate patterns. In this area the West Germans have a lead with their synchrotron X-ray source BESSY, but the Japanese are closing in. The facilities and skills of both countries should become available to American chip producers. Even giants like IBM have expressed concern that silicon chip suppliers, working independently of the microelectronics conglomerate, may be forced out of the market.

Some of the traits that launched Silicon Valley are now derided. Today the flexible young man who leaves his company to start a new business is chided. The managers in and around Palo Alto, once renegades themselves, now complain about the intolerable loss of time and skill when an employee leaves. The faithful Japanese technician, loyal to his company for a lifetime, has, strangely enough, become an ideal in Silicon Valley!

Japan has its difficulties also. The rise in the value of the yen relative to the dollar has necessitated some painful adjustments. Many worried managers confess to losses in the silicon business. They face a growing threat from South Korea, a major competitor with fierce national motivation. In spite of huge debts, Korean industry and government pour money into the silicon industry. The national goal is to capture 10 to 15 percent of the world market for memory chips. Koreans feel miffed that semiconductor sales statistics typically list their country under the heading ROW, meaning rest of world, after the columns for the United States, Japan, and Europe. Their efforts may make them number three in the world, and push Europe into the ROW category. A new crisis for semiconductors is thus shaping up in Asia.

10

Europe on the Sidelines

DESPITE its impressive history of scientific research, Europe plays an almost insignificant role in today's chip technology. European products in general compete very successfully on the world markets, and foreign trade is active—but not in microelectronics. The statistics are revealing. Per person and per year, people in the United States, Japan, and the European Community all contribute roughly equally to their gross national products, about $25,000 annually; the amount varies with fluctuations in the exchange rates, but it is clear that the situation is not very different on the three continents. When one compares the value of per capita production of silicon chips, however, the United States and Japan hold a clear lead over Europe: roughly twenty-five dollars annually per person against only about three dollars per person in Western Europe! Other economic comparisons indicate similar weaknesses. The market for semiconductor chips can be taken as an indicator of the production of modern equipment. Europe's market in 1985 was just barely three and a half billion dollars per year, having grown by less than 50 percent in the three years preceding (Fig. 8). Japan, with less than a third of the population of the European Economic Community, on the other hand, had a market for chips amounting to almost seven billion dollars per annum, having almost doubled in the three years preceding 1985.

The newer the semiconductor product, the more Europe depends on imports or on licenses for production technology, increasingly from Japanese companies. Germany and most of the other European countries have to import more than two-thirds of their integrated circuits. Most of these imported chips still come from the United States, but the imports from the Far East are growing rapidly, especially in memory chips, in which by 1985 the Japanese had solidified their monopoly.

During the seventies and eighties, this situation in Europe aroused neither public attention nor political action. The total world trade in chips was so small as to be forgotten by the statisticians of the "Eurocracy" in Brussels. Only a few of the big European industrial companies acted. The Philips Corporation, with headquarters at Eindhoven, Netherlands, acquired Signetics, in the center of Silicon Valley. They heavily invested in silicon facilities, including large new MOS plants in Albuquerque, New Mexico. This strong American position put the Dutch company on the charts of the leading silicon chip producers worldwide. Similarly, the French company Schlumberger became internationally important in chips.

In 1982 West Germany imported more than 76 percent of its office

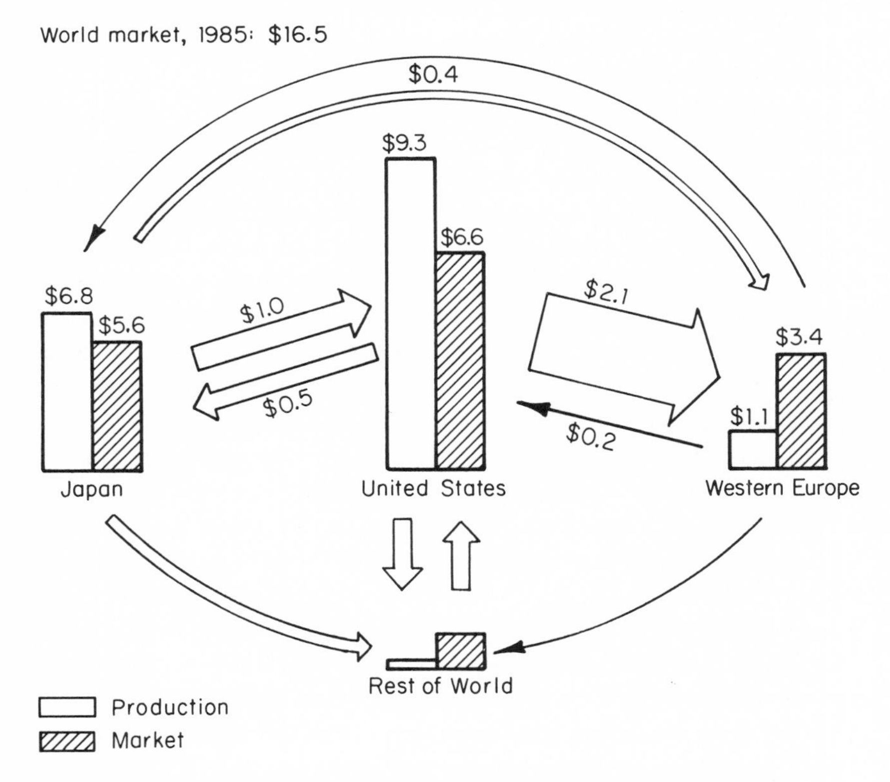

Figure 8. World balance of trade in semiconductor devices (not including the Soviet Union and China). The United States and Japan maintain export surpluses. Western Europe, by contrast, is increasingly dependent on imported silicon chips, even though the per capita consumption of electronics is lower than in Japan and the United States. Figures are from 1985, in billions of dollars.

machines and electronic data processing equipment. In 1983 imports in some of these areas had risen by more than 40 percent, while exports rose by only about 5 percent. Electronics is the growth industry of the future in Europe, as elsewhere. Computer sales in West Germany, estimated to be at least 5 billion dollars per year, are expected to achieve double-digit growth rates, according to the Central Electrotechnical Industry Association in Frankfurt. Yet a glance at the display windows of any computer or radio store reveals that the electronic products sold in Germany come from other countries.

As late as 1973, the currents of trade between West Germany and Nippon were fairly balanced, amounting to about 1.2 billion German marks. But in 1982 Japan exported more than 5 billion marks' worth of goods to West Germany, more than twice the amount that West Germany sold in Japan. Japanese microelectronics in computers and video recorders, stereos, and Walkmen flooded the German market. This trade imbalance may lead to more restrictive import barriers.

In July 1980 many West German newspapers carried an advertising supplement of several pages, paid for by Japanese firms. The shocking title read: "Japanese Electronics Conquers the German Market." At the end of the market analysis were the following words of a Japanese expert: "The Federal Republic of Germany really does lag behind the USA and Japan. It is even said that West Germany has given up this sector. But it is also known that Germany is far ahead of Japan in the research and development of nuclear energy for peaceful purposes, and can keep pace with the United States in this sector." To show that he meant well, the Far East friend then advised the German people: "You must begin developing new technologies!" At the time, Germans were hesitant to agree, but today the situation is different. Writing in *Siemens* magazine, Bernhard Plettner, chairman of the board of the Siemens Company, West Germany's largest electronic concern, praised the strengths of Europeans but noted, "We Europeans are weak in microelectronics and data processing."

Thriving branches of industry in Europe—including watches and cameras—have been wiped out or threatened by microelectronics. In June 1984, an American newsmagazine cover featured a cartoon in which, high upon a summit, Uncle Sam and a Japanese triumphantly shoulder an integrated circuit, while down in the valley below an old-fashioned European drags a mechanical clock to market. The British *Financial Times* warned that the Germans are in danger of missing the boat to the next industrial revolution and showed a cartoon of a desperate engineer holding his head in his hands. The caption reads, "Too often, they try to reinvent the wheel." German rigidity and exaggerated perfectionism are under attack. In Silicon Valley the daily *San Jose Mercury* stated that the German microelectronics scene is so inflexible and uncreative that the agile creators of the Apple computer would never have made it there.

International scientific journals reporting on new developments from European—especially German—firms find few scoops. Research in those companies' laboratories concentrates on catching up to the work of the Americans and the Japanese, who are at least two years ahead. Fewer patents and scientific articles originate in the Federal Republic of Germany. Japanese commentators no longer even consider Europeans as serious competitors. Europe has fallen by the wayside. Even newcomer Korea has in recent years invested much more heavily than Europe in modern plant equipment for semiconductor chips.

Only very recently have Europeans become aware of the significance of the silicon crystal for modern technology. The recurring headlines of the chip wars, the antidumping legislation, and the American debates have at last alerted the public. The original German edition of this book, which set forth Europe's comparative position in some detail, might have contributed to public awareness. A large development program called the MEGA project was initiated jointly by the Dutch Philips company, with headquarters in Eindhoven and wafer fabs in Nijmwegen and Hamburg, and the German Siemens Corporation, which has its headquarters in Munich and silicon fabs in Villach, Austria, and Regensburg, Germany. The Dutch and German governments supported the development of four-megabit dynamic memories at Siemens and one-megabit static memories at Philips. About 200 million dollars in government assistance helped to overcome the hesitation of the two companies to commit several billion marks and guilders to such a high-risk venture in a field where the Japanese were already so far ahead. In 1987 the first successful demonstrations of these two types of memory chips signaled some success in catching up. But the road ahead is arduous; Europe waited too long.

A Clever Policy?

Just how did Europe fall so far behind that many observers believe the Old World will never catch up again? Has the Continent become tired, dumb, or lazy? Who was sleeping, or was it even sabotage? It is easy to find polemic answers to these questions.

Strangely enough, it appears that even though it has ignored these new technologies, the Federal Republic of Germany has been able to chalk up quite a surplus in its balance of foreign trade. The United States envies Germany's situation; only Japan and the oil-exporting nations enjoy a similar trade surplus. Because only the most sophisticated markets can absorb large quantities of microelectronic goods, much of the world is excluded. The world still depends on West German suppliers of conventional products: motors, lightbulbs, and especially, first-class cars. Profits can still be made in these areas with new production methods, sound financing through government loans, flexible international sales policies, and good customer service. This reliance on conventional products, which led to Germany's postwar economic miracle, is still a viable strategy.

Many Japanese and Americans assume that West Germany is using the new technologies for defense projects and space programs, but the Federal Republic has no large-scale military or space projects that would require the new technologies. Unlike America, where the glamorous defense and space industries have drawn away many talented engineers and scientists, Germany still has a good supply of engineers and technologists for conventional fields such as mechanical engineering. Their fresh knowledge has modernized production in sectors such as textiles, furniture, and shoes, which are much weakened in the United States.

In Europe the long-standing markets for radio and color television have boosted the family of analog chips. Integrated circuits, retaining their analog functions, arose out of the many individual transistors on the television's printed circuit board. Unlike televisions and radios, data processing, the field in which Europe generally lags, requires digital circuits operating on the yes-no principle. Weak in most fields of microelectronics, Europe remains strong in power transistor electronics used in heavy machinery. The Old World has long dominated this branch of industry.

Is Europe pursuing a special strategy of outfoxing their allies in microelectronics? Is it perhaps watching from the sidelines as the two technology giants, Japan and the United States, fight each other to the death? If a firm can obtain cheap chips and build them into its own products, it will save money on research and development and reap profits on the entire system. It is not necessary to join the cut-throat struggle for price cuts, brand-new products, and markets. A top manager of a German firm that sells home entertainment and telephone equipment once set forth this strategy; he stated that chips should be considered raw material to be bought and refined. Statistics appear to support his claim. Silicon chips account for a maximum of 8 percent of the production cost of electronic devices. In West Germany, microelectronics makes up only a sliver of the total gross national product pie. So why all the commotion?

The commotion *is* justified, because that sliver of the GNP is much more important than it appears. Microelectronics may not seem to account for much in trade statistics, but it is being used more and more in machine tools, motor vehicles, measuring devices and instruments, telephones, and home entertainment products. More and more branches of industry will become dependent on the crystal. Any national economy that does not develop these products will be outdated.

In addition, Europe cannot disregard the fact that silicon technology has great potential for growth and expansion, offering new, high-quality jobs. In the United States electronics has become the industrial sector with the largest number of employees, well over 2.5 million in 1985, and far ahead of all other segments of the U.S. economy! Moreover, the information industry, with silicon as its shining star, is beautifully suited for nations with little space, and few natural resources and energy sources. Why, then, did Europe so drastically underestimate its significance? How could the Old World passively stand by as America and Japan took the lead?

Reasons for the Gap

There are three reasons why Europe, and especially West Germany, fell so far behind: the significance of silicon was underestimated, the economic miracle was based on conventional industry, and public and government priorities lay elsewhere. In the beginning many people underestimated the possibilities of the silicon crystal; even the experts never dreamed of its myriad uses today. Cooperative ventures by American and European firms were confined to large-scale projects such as nuclear energy. The information flow from the firms in Silicon Valley was insignificant by comparison. For various reasons, small American businesses did not gain a foothold in Europe.

The postwar reconstruction of Europe's factories and export-oriented traditional industries required huge investments of capital and labor. Germany's profit margin was dangerously lower than America's, and its social benefit costs were higher. Electronics looked too risky. Economic planners figured that since there was no major military market for the new technology, and chips could always be bought inexpensively from Silicon Valley, they did not have to invest heavily in expensive silicon research.

Europeans were relatively slow to realize the importance of computer technology. Large-scale international data processing projects were set up by various governments in the sixties, with the aim of creating a unified European computer industry to compete with IBM, but they all failed. Government bureaucracy had no chance on the free market competing against successful multinationals. Japan could afford to compete against American data processing firms, but NATO-allied Europe could not. Despite great efforts by its politicians, Europe was unable to create its own digital industry.

The U.S. military sponsored a lot of basic research at American universities. Indeed, the close cooperation between the Pentagon and academia was often criticized. But no such links were forged on the Continent. Germans in particular found such collaboration unthinkable in view of the past two wars. By tradition, university professors maintained their independence from princes, generals, and government officials. Because of these attitudes, governments are reluctant to spend even a fraction of defense funds on academic research and education in new technologies.

Regardless of the warnings, Europeans had not significantly changed their attitudes toward microelectronics by the early eighties. As a result, traditional industries watched their markets dwindle and competition from the threshold countries develop. Unemployment, an almost forgotten phenomenon, suddenly broke out anew in Europe. Those affected cried out for new fields of employment, but most were already occupied by the Pacific Rim competitors. And to make matters worse for the Old World, new silicon-processing methods force large-scale automation, which further aggravates the social problem of unemployment. By importing microprocessors, Europe is exporting jobs.

Bruce Nussbaum's biting, controversial book, *The World after Oil*, published in 1983, ignited discussion of Europe's pitiable role in the battle for new technologies. Nussbaum attacked the old smokestack industries for their inability to regenerate themselves, especially in America. Describing the merciless trade war with Japan, he warned that today's decisions for microelectronics will affect the next century. Nussbaum cautioned his American readers to beware: the Federal Republic of Germany is to him the most terrifying example of an industrial nation on the decline. *The World after Oil* pictures West Germany as crippled by socialism and bureaucracy, age and *angst*. Nussbaum joins other critics in saying that the key factor in the decline is the country's inability to master microelectronics. Nussbaum feels that West Germany will be forced to cooperate with the Eastern Bloc, which relies on what he depicts as incapable, oppressed, and continually inebriated workers. Instead of taking advantage of labor opportunities next door, West Germany continues to bank on conventional industries such as steel and concrete. Indeed, jobs in agriculture and coal mining increased by 6 percent from 1974 to 1984, while jobs in the computer industry fell by 25 percent.

"If ever there was an economic environment made to discourage entrepreneurial activity," Nussbaum laments at the end of one chapter, "it is now Germany. If ever there was a nation poised to stumble as the world moves into the post-OPEC era, it is Germany." The book's German edition was much tamer; interviews were added to soften its unpleasant impact. Some of his warnings, however, led to parliamentary hearings, public awareness, and political activity. The media clamored for microelectronics experts, and critics pounced upon each new evidence of economic decline, such as the demise of the television manufacturer Grundig or the poor performance and final divestiture of Volkswagen's affiliate, Triumph-Adler.

Other American observers, however, take a different view. Peter Drucker, for example, in the *Wall Street Journal* (March 6, 1986), praises Germany's economic policy for the broad base of its products, the absence of strong dependence in either exports or imports, and the concentrated effort to maintain a high level of competence in conventional products. A moderate supply-side economic policy and no plans for large-scale government-sponsored research are positive signs to Drucker; the lag in microelectronics does not appear as a great setback.

Europe in the Silicon Age

What has Europe been doing? This cluster of loosely allied, somewhat unified nations is primarily concerned with agricultural problems. Its ministers discuss such issues as production of olive oil and milk, price adjustments, tariffs, and tax levies. They do not spend much time planning joint efforts in new technologies. The poorer members of the European Community are even less interested in high technology than the rich ones. Ireland, for example, is fighting to gain additional percentage points for its milk quota. At the same time, it plays up its attractively low wage structure to appeal to American electronics firms.

Some figures are telling: in 1980 the European Community pumped forty times more money into agriculture than into research and development! The Economic Community nations invest roughly five billion dollars in subsidies for dairy products. Even a tiny fraction, say one percent of this amount, would be a major contribution in a joint developmental program for European microelectronics.

Several European nations have joined in creating a research program in communications technologies, known as ESPRIT, but such cooperation is rare. Any European microelectronics program will have to consider the interests of all the participating nations, which are too often at odds with each other. The different levels and requirements of technology in each country make agreement difficult. First of all, England and France have long-standing military interests that ensure government support for the most modern electronics. The United Kingdom has uninterruptedly promoted research and development in this field, even though their efforts have not led to a big breakthrough. Compared to the rest of Europe, microelectronics in Great Britain is relatively strong.

For a while, one out of three European chips was made in Scotland. The area between Glasgow and Edinburgh had become a new "Silicon Glen." The low wages in this economically depressed region and the availability of good workers attract international firms, offshoots from Silicon Valley and Japan. Semiconductor firms are able to take advantage of support from Heriot-Watt University in Edinburgh, very inexpensive property, and generous tax breaks. For Americans it is faster to fly to Scotland than to Asia, and, of course, everyone speaks English, an important consideration for an industry geared to complicated processes and directions. In Greenock and Queensferry diffusion furnaces process silicon wafers and factories produce electronic measuring devices. In Livingston a new facility for a Japanese silicon manufacturer, Shin-Etsu-Handotai, will polish crystal wafers, then send them to Japanese subsidiaries down the road for further processing. By 1984 five hundred firms provided several thousand jobs in Scotland. But the decline of the American mother companies in 1986 affected the Scottish daughters even more severely. The hopes for a European Silicon Glen have diminished sharply.

France has taken a different route. In the days of right-wing governments in the 1970s, France sought American partners, but this was followed by a great wave of industrial nationalization under the socialist administration. France's computer development sector and semiconductor industry had already been supported by large-scale state planning, including a *plan calcul* for computers and a *plan composants* for semiconductor devices. Nationalization led to larger agencies and a single monolithic industry. Despite its great losses, France's Thomson concern continued to buy major shares in other companies, most importantly, in West Germany's traditional Telefunken corporation. In doing so, the French not only improved their market opportunities for televisions and radios, they also expanded their market, now regulated, for silicon chips. Buying into Telefunken as a chip consumer also broadened France's market base for new circuit developments, which the country's military program requires.

The French Postal Services are recognized as having the country's most modern telecommunications systems and being a vehicle of remarkable modernization. The "minitel" network of information and communication throughout France, based on heavily subsidized, low-priced television monitors, was a remarkable success for modern microelectronics in telecommunications.

The large new laboratories in Grenoble are becoming a center of electronics technology. France is taking careful note of Japanese examples. The government-aided but private French takeover of the American International Telephone and Telegraph, ITT, in 1986 may indeed be the beginning of a giant worldwide telecommunications group under French leadership.

Nations south of the Alps follow yet another strategy. SGS-Ates, an Italian firm previously plagued with chronic deficits, is now more energetic, aggressive, and successful. Its head, Pasquale Pistorio, worked in one of the big silicon firms in America before returning to Milan, where he signed contracts with Japan and Silicon Valley. Because of this and other Italian successes, *Time* magazine reported in March 1984 that industrial history was once again being made in Italy. But the Olivetti firm received a welcome financial boost when AT&T bought a quarter of a billion dollars' worth of its shares. Their venture marks the start of highly significant cooperation between Europe and the American telephone giant. Through Olivetti, U.S. silicon processing will reach Europe.

The Netherlands probably enjoys the strongest position of the European nations, because Eindhoven is the headquarters of Philips, which is actively increasing its share of silicon technology. The firm has research and development, as well as production ventures, in Japan, Silicon Valley, and other parts of the United States.

This cursory survey of Europe's microelectronics industry makes it clear that because of the significant differences in development among its nations, Europe will not easily follow Japan's example of a unified industrial strategy. Nor will it easily adopt the American model of very large and very small high-technology firms strengthened by a unified national policy. At this point European Community cooperation appears limited to establishing protective import barriers. Europe's nations compete, not with the United States and Japan, but with each other in wooing American and Japanese firms. As a result, the strongest thrust for unifying scientific and technological efforts in Europe is provided by the multinationals: IBM, ITT, Hewlett-Packard, and Intel, among others.

How will Europe fit into the world economy? Some cynics offer this prognosis: Nippon will act as the factory, supplying the latest equipment with silicon brains and nerves; the United States will supply raw materials; and Europe will offer culture, castles, and couture to sightseers. Exaggerated for shock effect, this prognosis nevertheless points up the serious and growing technology gap between Europe and the rest of the world.

Germany and the New Age

In early 1984 the consultant firm Gnostic Research presented a graph of information technology production values. The United States showed a phenomenal increase of more than 100 billion dollars annually from 1975 on. In 1966 Japan and West Germany were about equivalent in this sector, but by 1982 Japan had become almost three times as strong and was gaining. In Germany exports in the electronic data processing category sank from 40 percent in 1972 to a mere 14 percent in 1983. In the telecommunications industry, exports dropped from 23 percent to 14 percent. During the same period Japan and the United States made huge gains on these world markets.

Was Nussbaum's stinging critique justified after all? Research and development of new technologies are much less unified in West Germany than in Japan and the United States. Japan's MITI is directly answerable to the prime minister, and the U.S. Department of Defense finances a large portion of university research in new technology. The president has a science advisor in the White House.

The chancellor of the Federal Republic has no such direct role, although the minister of economics does have a science and technology research institute similar to Japan's MITI. But this agency's tasks are basically to set standards and norms. The organization does not conduct microelectronics research and development and has few funds for basic research. The German government must exercise caution in promoting research because of the risk of infringing monopoly laws. Indirect tax breaks are acceptable, but direct grants to industry would run counter to free-market theory, which claims that if microelectronics is really so important, market forces will ensure it a primary role.

West Germany's telephone system, part of the Federal Postal Services, could, like Japan's Nippon Telegraph and Telephone, assume a central role. But even though the telephone system has the funds to finance research and development, it must surrender any profits to the Federal Ministry of Finance. Moreover, its very good research lab is much too small to spearhead future technology.

Public opinion in Europe concerning research and development for microelectronics is not exactly positive. The government-assisted Dutch-German MEGA silicon memory chip project is more criticized than applauded. Although much larger amounts of tax money are being poured into the dying steel industries in the Ruhr and the Saar with little criticism, support for MEGA has been heavily attacked by both the left and the right. Most important, the individual jobless miner seemed to be pitted against the rich electronics companies. Had not Nussbaum predicted this type of sentiment? The outcry became outrage when Siemens announced that they had taken a license from the Japanese microelectronics company Toshiba in order to catch up quickly in the manufacture of highly integrated memory circuits. It looked as if the government subsidy had been accepted just to be passed along to Japan. Nobody really understood the significance of time in the competition for this novel business, where the learning curve of price decay wields its merciless whip.

The unusual boom of 1984 in the silicon chip market revealed the strategic importance of access to chips. When the world suddenly clamored for silicon to the point that suppliers could not keep up with demand, Germany's conveyor belts came to a standstill All available devices went right to the markets in the United States and Japan. Like other European nations, West Germany is vulnerable to any supply bottleneck. Conceivably, the United States could put an embargo on chips for military reasons, or Japan might decide to sell only finished products. In any case, German industry will suffer.

Social scientists will eventually explore why Europe has been so slow to adopt, and even been downright hostile toward, modern microelectronics. One reason is the poor business climate for small, risky ventures. Unlike American entrepreneurs, many of whom can boast of success stories, Europeans who have tried to enter the microelectronics field have largely failed. Another reason is Europe's general resistance to change. Even Germany's dynamic, new Green Party advised their candidates to oppose "inhuman" microelectronics. Contributing to the resistance is a widespread fear of the loss of jobs. This sentiment was strengthened by the sudden surge of unemployment in the mid-seventies just when microelectronics was beginning to invade industry. Yet detailed studies of the slump showed that most of the jobs lost were in heavy industries such as steel, shipbuilding, construction, and textiles. The introduction of the microprocessor had little effect on employment. At an international conference ominously entitled "1984—The Orwell Year—and After," American observers gave detailed proof of the lack of correlation between microelectronics and unemployment. At a meeting in 1985, former "Eurocrat" commissioner Count Davignon told a disbelieving socialist audience that Europe would have had about eight million new jobs if it had introduced the silicon crystal as fast as Japan and the United States had.

A cultural bias has hindered the development of microelectronics as well. Europe has traditionally considered businessmen and intellectuals as belonging to two distinct cultures. But the new technology requires a harmonious interplay of intellectual scientific curiosity and practical business that is difficult to find in Europe. Proof of this came recently when a German chemical company could not find a European partner of sufficient stature to do basic research. The firm subsequently granted the largest research contract in history to Harvard University and Massachusetts General Hospital. This split between business and research may be as much of a problem as the tax laws in most European countries, which do not favor venture capital and penalize capital gains.

High customs barriers that are supposed to protect manufacturers will actually have the opposite effect. Integrated circuits are already more expensive in Europe because of the import duties levied—up to 17 percent. As a result of this high import duty on the most important component of electronic systems, European manufacturing and export of the systems suffer. To stay competitive, European manufacturers will be forced to set up factories in the Far East or America, but that will cause further outcries at home. Short-term protective policies are not the answer; Germany, like the rest of Europe, has to take control of its technology. A personal computer has two special keys on its keyboard. You hit the "return" key when you want your program to stay alive. The other key, "escape," you press when you want to get out. Europe can still choose which key to press, but it is getting late.

It appears that Europe's future contribution to the skillful art of semiconductors will be modest, but the situation is not hopeless. Further erosion of its market share may be avoided, and even a little gain seems possible for the last decade of this century. Possibilities for international cooperation offer some optimism. The weakening of the American position, especially for the smaller silicon companies, created a more advantageous position for the big European electronics concerns with their in-house markets and their financial strength. Investment for each new line of chips has grown to such proportions that the large, vertically integrated systems companies have a much better chance than the small, flexible specialists, such as those in Silicon Valley, which dictated the technology and the speed of development in the sixties.

Two blocs of silicon chipmakers may emerge in Europe. One would be the already tightly cooperating union of Philips and Siemens, each having their own international liaisons and affiliates. A counterpoint to this Germanic bloc might be the Latin confederation of the Italian SGS and the French giant Thomson. Their chief executive officer, Pasquale Pistorio, who had many years of experience in Arizona, seems capable of merging the French and the Italian segments into one powerful organization. They own the big wafer fabs in Colorado and Texas that were once part of the American Mostek Company, a world leader in memory chips before it went bankrupt and was taken over by a French company. Pistorio says, "We Europeans must achieve our comeback, we just need a strong semiconductor industry . . . but the Americans will also come back, their society cannot tolerate abandoning their leadership."

11
The Crystal Microcosm

. . . .

We are approaching the end of the Iron Age. It is not that metals have outlived their uses; it is simply that silicon, with its seemingly infinite uses, has outpaced all preceding material technologies. The coming era, which will be based on scientific methods, is the Age of Silicon. Trial and error, careful observation, learning from failures, and a kind of intuition for technical possibilities—all of these time-honored methods are no longer enough when working with the sensitive semiconducting crystal. Only the scientific understanding of how atoms interact within the crystal lattice has permitted the development of modern electronics. The new technologies require precise diagnoses, numerical checks, mathematical discipline, and an incredible degree of purity. Tactile and visual observations no longer serve the researcher; the functions of a semiconductor are hidden in the spatial arrangement of atoms imperceptible to the human senses.

In 1972 the number of scientific publications on the properties of iron and of silicon were about equal. In the next ten years the number of publications on silicon rose from 500 to 3,300, or about ten articles a day! In contrast 700 works on iron were published. No other substance, not even water, has been so thoroughly investigated as silicon. Its composition, atomic structure, deviations, reactions to light, magnetic fields, interfaces with related materials, electrical conductivity, and other properties are known in more detail than those of any other material.

The eventual triumph of silicon over thousands of other semiconducting materials was not a foregone conclusion. Many other elements possess properties that are even more appropriate for certain applications. But silicon's structural stability, simple chemical makeup, and high degree of attainable purity made it valuable. In addition, the crystal's versatility could be enlarged by doping it with electron-donating and electron-accepting atoms. But its most important property is the unique ability of its oxide to combine with it and provide a protective covering. The oxide skin of a semiconductor chip acts as a protection, a base for patterns, and a mask for processing. Finally, the price of mass-produced silicon chips is so low that other materials can hardly compete.

It is fascinating for an expert to watch a new technological system develop optimal solutions while branching out and accelerating its own evolution. The layperson has trouble appreciating the fine dimensions possible in today's silicon crystals. The following example may help: if you split a human hair vertically, you would find in the cross section along the cleavage no less than 100 memory cells of the size of the next generation of memory chips! Just how carefully must the crystal's surface be prepared? The following comparison tells the story: an irregularity equivalent in size to a grain of wheat in an area as large as forty football fields is the limit!

Lighting the Way

In microelectronics silicon truly holds a place in the sun—and not only figuratively speaking, for the *p-n* junction in a silicon crystal can transform sunlight directly into energy. When a light quantum, called a photon, from the sun enters the crystal, it forces one electron out of the structure of linked atoms. The remaining hole acts like a positively charged particle. An electrical field is always present at the junction between an *n*-zone and a *p*-zone. This field, created by the crystal's structure and by the dopants introduced for that purpose, draws the electron and the hole off in separate directions. Positive holes migrate to the *p*-zone, while the negative electrons stream to the *n*-zone. The resulting separation of charges produces electric current just as it does in a battery or generator.

This direct transformation of light into electrical energy uses no moving parts and is noiseless, odorless, simple, and environmentally safe. Although it is necessary to erect large, flat, often ugly black structures for solar energy systems, these could be economical in sunny equatorial regions or in places far removed from other power sources—

such as in the high mountains, on buoys at sea, or in desert oases. But obtaining large amounts of energy from silicon is too expensive generally to compete with coal or nuclear energy. On the other hand, small amounts of energy can be generated cheaply with this technology. Pocket calculators made with energy-saving silicon chips, for example, operate on daylight alone. Japanese manufacturers, which had already conquered the market in pocket calculators, were the only ones to recognize the opportunity presented by the new devices.

Silicon solar cells, which already power many tiny Japanese computers, don't require symmetrically ordered, expensive crystals. Instead, they use amorphous silicon, in which the atoms are arranged irregularly, thus lowering the energy yield but reducing the cost. Instead of being painstakingly drawn out of the hot crucible, the material is quickly evaporated onto a large surface. The silicon crystal has thus created its own competition. A few years ago some people caused a commotion by claiming that amorphous silicon would also fulfill the demands of electronic circuits. But the requirements of precision and high yield per unit demand the symmetry of crystalline form. Only when silicon is used to transform light into energy are symmetry and exact control not necessary.

The battle of the silicons is not yet over. It is still not clear which form will be used in the future to change sunlight into electric current on a large scale. A compromise may be possible, however. Instead of large, flawless, perfectly symmetrical single crystals, a mixture of the finest crystallites, called polycrystals, might also be used. Efficient solar cells made of polycrystalline material are already being produced in large quantities. A solar-cell energy program of the West German government has paid off—the German Heliotronic Company has gained a leading position in the field. The next step is to come up with an inexpensive mass production process. In the United States, another approach was strongly supported during President Carter's administration: amorphous, completely noncrystalline, silicon made by deposition of vapor in very thin layers can be used for less efficient but less expensive large-area solar cells.

Silicon's Competitors

Although silicon is a remarkable material, it does have one limitation: light cannot be produced inside its crystal structure; the transformation of light into current via the solar cell cannot be reversed. Production of light from electric current is important; for instance, a small, strong, efficient source of light is needed in telephone technology using glass fibers. Light bulbs are too expensive and awkward. In this area, a new family of semiconducting crystals is outdoing silicon: gallium arsenide and its relatives can directly transform electric current into light—and can even generate laser light.

In the fifties, Heinrich Welker and his research team at Siemens discovered the properties of these crystals. In trying to comprehend the semiconducting characteristics of germanium, Welker studied the properties of individual atoms. Instead of two germanium atoms, with their total of eight outer electrons, Welker used two different atoms that also hold a total of eight outer electrons. He chose the two elements on either side of germanium in the periodic table: arsenic, with five outer electrons, and gallium, with three. Together they form gallium arsenide. The two elements with their differing electrical charges create a polar semiconductor crystal; no longer an elemental semiconductor like silicon, it represents the family of compound semiconductors, often abbreviated as "three/five compounds" because gallium is in column three and arsenic in column five of the periodic table. Gallium arsenide differs from germanium and silicon in two important ways: the electrons are more mobile and traverse the crystals more rapidly, and they couple more strongly with light.

The high speed of the electrons in gallium arsenide gave rise to great hopes. Although the Siemens scientists worked feverishly on technological discoveries, their practical applications were not successful. Despite their many advantages, the new crystals could not outdo the simpler, more reliable silicon. The gallium arsenide setback led to greater caution in both research and business.

After years of massive government-subsidized efforts in this area, success finally came to the German and Dutch labs of the Philips Company and then to the United States. Clean crystals of gallium phosphide and gallium arsenide were grown, and many different colors of light were coaxed from them. Inside these crystals, electrical current washes the electrons from *n*-type regions into zones with many holes. This well known *p-n* junction unites electrons with their antiparticles, the holes. When the electrons jump into the vacant space, they must release energy, which they do by emitting light. By doping gallium arsenide with silicon, scientists achieved a remarkably high yield of invisible infrared light out of the gallium arsenide crystal. Silicon, with its four outer electrons, can sit on a gallium site as well as on a site reserved for arsenic atoms within the gallium arsenide crystal lattice. Silicon thus has a dual function when introduced as a dopant impurity into gallium arsenide: it acts as a donor on a gallium site and as an acceptor when replacing an arsenic atom. This duality proves useful for the conversion of electricity into light.

Today, these semiconductor crystal light sources are used in indicator lights in all sorts of products, such as television remote-control devices. But that is not all: gallium arsenide can also sustain a sharply focused light emission that can be shaped into a laser light source. When electric current is directed into the crystal, tiny bits of finely tailored crystals create light that oscillates in a strictly synchronous manner, fulfilling the coherence requirement of a laser beam. Light oscillates much more rapidly than radio waves and can thus be encoded with much more information. A laser diode beam, visible only with a microscope, can also scan the information encoded in the minute perforations of a video disc. No direct contact takes place; only a beam of light scans the tiny holes. What an improvement over scratchy old phonograph needles!

The new optoelectronics are made possible only by compound semiconductors such as gallium arsenide and gallium phosphide. Crystals composed of just one kind of atom, such as silicon and germanium, cannot produce light from electric current. Optoelectronics rejuvenated the fields of atomic physics and chemistry. New processes for manufacturing strictly controlled crystal layers have been created. In highly evacuated vessels, researchers direct a carefully controlled stream of atoms onto a crystalline base or substrate to build up, layer by atomic layer, a new crystal. Following precise stipulations, the researchers can tailor the composition of the crystal to meet specific requirements. A laser, for instance, should emit light in the color least likely to be absorbed by the conducting glass fiber. Compound semiconductors can be mixed to produce any desired property.

As an information processor, silicon faced a second serious rival in the 1980s—magnetic storage. The magnetic storage of information in a crystal was termed magnetic "bubble" memory. By this method, "yes" and "no" functions were produced through the north or south poles of a magnetized zone, or bubble. Using garnet crystals with admixtures of magnetized atoms, the technique promised the possibility of manufacturing densely packed information memory chips. Magnetization was both economical and efficient. Long-lasting, it did not require continual restrengthening with current impulses, as did the silicon memory chip; such a storage principle is thus called "nonvolatile." In the past few years rivals Bell and IBM have spent huge sums on research on this nonvolatile storage principle. But silicon survived with Darwinian ruthlessness. As its structures were mastered, prices fell, and the scope of its applications became enormous, pushing the magnetic competitor aside.

Confined Quantum States

This technique of molecular beam epitaxy has triggered basic research using even finer materials. The layers of today's crystals are so thin that the tiny spaces into which the electrons are forced evince quantum characteristics. The crystal has thus become a laboratory for quantum theory experiments. One of the most exciting developments was not produced in the mile-long, expensive tunnels of high-energy physics research facilities; it took place right inside the crystal.

Other investigations into the properties of semiconductor atoms have led to radical discoveries. A German solid-state physicist, Klaus von Klitzing, working in the joint French-German laboratory for strong magnetic fields in Grenoble, decided to investigate electrons on a semiconductor surface in a magnetic field. The Grenoble facility, operated by the Max Planck Society and its French sister organization, the C.N.R.S., is accessible to all universities. Von Klitzing asked his friends at the Siemens lab for a sophisticated silicon field-effect transistor for an experiment. He found that when electron motion is doubly restricted, by the electric field near the crystal's surface and by the magnetic field forcing the electrons into circular orbits, quantum forces become dominant. All other factors, such as length and width of the conducting channel, temperature, and amount of doping, are inconsequential for the magnitude of resistance exhibited. This phenomenon, called Hall resistance, is determined by two constants: Planck's action quantum h and the quantum e of the elementary charge of the electron.

The "von Klitzing effect" is a milestone in our understanding of basic principles of solids, unhampered by such accidental properties as chemical composition and geometrical dimensions. Chemical purity and crystalline perfection are necessary, of course, but in the interpretation of the experiment they disappear. This situation is exactly opposite that of a century ago, when the crystal's fickle irregularity obscured all physical laws. No wonder von Klitzing won the Nobel Prize for physics in 1985.

Europeans suddenly learned about the scientific significance of semiconductors when the sensation of a Nobel Prize after so many years excited interest in solid-state physics. American researchers, especially German-born Horst Stoermer and Chinese-born Dan Tsui, not only repeated von Klitzing's experiment, but they and others discovered a complicated fine structure of this Hall resistance, termed the "fractionally quantized Hall effect." Subtle interactions between the electrons become visible at very low temperatures and very high magnetic fields; the theory is not yet completed and must be refined to adequately interpret these intricacies, which now can be observed for electrons inside a semiconductor crystal.

Tunnels and Superconductors

Research into crystals was pushing ahead. Scientists in Zurich, the headquarters of IBM's European basic research, succeeded in developing a "tunnel microscope" capable of scanning the surface of a crystal. Ferdinand Braun's old technique came back at an astounding level of sophistication. Two physicists, the German Gerhard Binnig and the Swiss Heinz Rohrer, at the IBM Research Laboratory at Rüschlikon, near Zurich, constructed a new type of microscope with exceedingly fine resolution. The technique is based on the tunneling of electrons over very short distances. In tunneling, a property of particles predicted by quantum theory, a particle penetrates an energy barrier that would be considered impenetrable in classical physics. Rather than surmounting the barrier, which would require a lot of energy, the particle can "tunnel" through a barrier if it is only a few atomic distances in width. Very slight variations in this distance cause rather large variations in the tunneling probability. The tunneling electrons can be detected by a measurement of the current, whose variations can thus be translated into measurements of distance between the needle and the crystal surface. With this technique it is possible to obtain a very sensitive picture of the arrangement of atoms on the crystal surface; where the plane surface terminates abruptly, the atoms are arranged in puckered patterns. The tunnel microscope was greeted with great enthusiasm, and Binnig and Rohrer were awarded the Nobel Prize for physics in 1986.

Understanding the crystal surface had always played a vital role in developing technical utilization. One major question concerning the future of microelectronics is whether the increasingly skilled exploitation of surface areas will lead to the mastery of space. Scientists conceive of densely packed three-dimensional chips. They need only learn how to connect or separate—according to the operation desired—the layer upon layer of active zones. Encouraging signs point to success. A new technique of constructing crystals by using molecular beams can create extremely thin layers. With depth under control, the two remaining dimensions can be shaped as desired by etching. How can the two techniques be united? First approaches to solving this problem are being made by computer. IBM has poured huge sums of money into research on the use of ceramic substrates to mount and connect silicon chips. Scientists have created three-dimensional ceramic bodies by stacking numerous layers, each with its individual pattern of connections and channels. Inside this structure are intertwined paths for currents moving between the memories and the logic devices. The connections and crossings of the many paths provide many more possibilities than are found on the surface. The third dimension is being regained.

Meanwhile, in a totally different branch of modern physics, another competitor hit the spotlight: superconductors. In the United States alone more than 100 million dollars was spent on research into superconductivity, led by IBM laboratories.

Every switching operation in the silicon crystal generates heat, which places limits on further reductions of scale in memory chips. When the semiconducting crystal becomes too hot, it loses those properties that enable it to differentiate between zero and one. This phenomenon, termed dissipation loss, stakes out the boundaries of the silicon crystal. Water and other cooling methods have expanded those bounds somewhat. But researchers wonder whether it would be possible to dispense with the silicon crystal altogether, replacing it with a piece of superconducting solid. Superconductivity refers to the fact that some metals at very low temperatures provide absolutely no resistance to the flow of an electron stream. At an international conference Brian Josephson, a young English researcher, postulated a strange theory. Two metal strips, when separated by a thin isolating layer, form a kind of switch, which can go rapidly back and forth between two states. The switch can be used for ones and zeros corresponding to yes-no alternatives in digital information. At first nobody believed that Josephson's concept could be correct, but his hunch turned out to be right. in fact, the theory was named for him, and he was awarded the Nobel Prize. The Josephson Element seems especially promising because it works at very low temperatures with materials that facilitate the transport of current and so requires a very tiny amount of energy. Because the energy-saving switch can perform elementary storage and switching operations at astonishing speeds, the Pentagon wanted it for a high-speed supercomputer. The Japanese followed each new development with great interest.

In the summer of 1983 anyone spending time in the corridors of the IBM lab in Yorktown Heights, New York, was sure to witness some heated conversations; the entire Josephson Project, employing about 120 scientists, had just been cancelled. Silicon had once again defeated a competitor. The cryogenics technique had proven too problematic; the metals couldn't endure the extremes of heating and cooling. But even more important, the metals had failed to meet the demand of amplification. A computer has to transmit signals over long distances. The signals weaken in the process and must be restrengthened, and here the superconductors could not offer a convincing solution.

But after IBM announced that it was abandoning the troublesome superconducting metals, a veritable sensation sent the solid state physics community into a frenzy. In January 1987, Karl Alex Müller, an experienced physicist and privileged IBM fellow, and his associate Johann Georg Bednorz, a young German mineralogist and physicist, working at the IBM Zurich Research Laboratory, announced that they had found a new class of solid-state materials that could become superconductors at much higher temperatures than any previously known metals, alloys, or compounds. Then began a feverish international race to find materials that would be superconductors at even more reasonable temperatures, with researchers finding new combinations of oxygen with barium, lanthanum, and other atoms. Such compounds are not really metals (which were previously thought to be the best superconductors); instead, the new crystals resemble oxide compound semiconductors with doping so heavy that they approach metallic conduction. Copper atoms, lined up regularly with oxygen atoms, seem to be the essential structural elements.

The hectic March 1987 meeting of the American Physical Society, held in the Hilton Hotel in New York, was later nicknamed the "Woodstock of Physics." At the time of the December 1 deadline for submission of papers to be delivered at the meeting, no one knew anything about the copper-containing perowskite crystals that were exhibiting superconductivity at the higher temperatures. When the news from Zurich became known, the meeting organizers frantically tried to add some last-minute talks on the subject to the program. As it turned out, the entire meeting was dominated by discussions of superconductivity, on into the night and the wee hours of the morning. What were the materials? Why did the phenomenon occur? How could the temperatures be raised even higher? For one evening session in one of the largest meeting rooms, I arrived an hour early to get a seat, but that was impossible. Solid state physicists were used to being looked down upon by their elementary particle colleagues, who commanded entire divisions in their giant laboratories and spent billions of dollars on studies that had no practical applications. The jubilation at the New York meeting sounded like the cries of the perennial underdog, finally victorious.

Never before has the Nobel Prize selection committee acted so swiftly and spontaneously as they did in 1987 in awarding the prize for physics to Müller and Bednorz, who had hardly been known outside of a small circle before their discovery.

This new breakthrough came as a total surprise to theorists. When superconductivity in these novel materials was found at temperatures as high as −173 degrees Celsius, it became clear that a massive revolution in electricity could be imminent; lossless transport of energy could be possible without expensive, complicated cooling systems. The creation of wires and cables will require a great deal of further technological development. IBM engineers have already demonstrated thin films with remarkably high current-carrying capacity, but much work remains to be done before it will be known whether superconductors can compete with silicon or perhaps combine with them in totally new ways.

Looking Ahead: Biomolecules

Laboratories around the world are exploring the third dimension. Crystal growers in particular are trying to predetermine the exact directions of the crystal's growth. But the real barrier to placing electronic functions inside the mineral is the heat that develops from every process of switching, amplifying, storing, and conducting of electrical charges. The unregulated, uncontrolled movement of heat compromises all atoms. Its expansion inside the crystal cannot be reversed. Traveling through the crystal, electrons repeatedly collide with atoms, to which they transfer the energy of their motion, thus heating the crystal. The heat problem has been lessened, but it must be entirely overcome if the crystal's interior is to be packed with electronic functions.

Anyone who tries to improve microelectronic operations soon appreciates the beauty of the structure of living cells; biological mechanisms use a minimal amount of energy and space. The physicist can only strive to imitate living forms with inorganic matter. Nature's cells and mankind's semiconductor devices do share some structural elements, especially at interfaces such as cell walls and *p-n* junctions. In both, electrical potentials are directly linked to the imbalances of ions and electrons at the junction or where cells meet. But the conduction and storage of information differ radically between biological and semiconducting systems. Nature is slow in the elementary steps because large molecules and ions are the vehicles rather than the rapidly conducting and switching silicon circuits. Yet nature compensates for the slowness by much tighter packing. Little waste heat is generated in our brain compared with the enormous and troublesome amounts of heat produced as an undesirable byproduct in semiconductors. Nature uses highly specialized molecular structures that adapt beautifully to the desired function. Semiconductors, by contrast, utilize increasingly dense arrays of units all doing the same basic function—at great speed, but often with an organization that appears unimaginative in comparison with biological systems.

The gate of an MOS silicon chip can be compared to the neuron of the human brain; both are the building blocks of their respective, incredibly different, systems. Today's gate measures about fifteen one-thousandths of a millimeter; a neuron, which is about three times that size, makes do with only a thousandth of the energy required by a silicon gate. In addition, in the course of its evolution, the organic version has gained the distinct advantage of "system architecture."

In the coming decades one big question will be whether inorganic systems will approximate the organic or move away even farther. Can the silicon atom hold its own against the multifaceted carbon atom?

The two atoms are related; both have fourfold symmetry in the electron clouds bonding them to neighboring atoms. The infinite variety of compounds in organic chemistry—from ordinary alcohol to the nucleic acids carrying biological information—is based on the properties of the carbon atom. More brittle and less varied is the silicon atom. Within the symmetrical lattice of a pure silicon crystal, a few electrically active, foreign atoms are introduced at the site of a silicon atom. As the electronic chips of the future are packed tighter and tighter, more and more foreign atoms will be introduced. Will these chips become as complicated as their natural counterparts? We do not yet know if such an assimilation of silicon to natural paradigms will occur. One thing is sure, however: the future of microelectronics lies not only in the primitive capabilities of the crystal's inorganic functions, but also in dramatic confrontations with the structures and functions of life itself.

Some hope for an entirely new approach to storing, switching, and manipulating information is pinned on the still rather vague idea of "biochips," a new class of devices that would be, it is hoped, better and faster. Such chips would be closer to a compromise between silicon and the organic structures of living organisms. An almost infinite number of organic chemical compounds can be synthesized from the known atomic building blocks. In principle it should be possible to build in a property of sensing external stimuli, such as light, the transport of energy and charge, and the storage and output of charge. Circuits of this type might be more densely packed, more specific and varied in their architecture, representing an advance over the monotonic regularity of the inorganic silicon crystal. But the great promises of a few years ago about biochips have thus far not materialized.

The possibility of transforming organic materials into fairly good conductors is also being studied. Scientists hope to create large organic molecules with clearly specified structural elements that can discriminate between a zero and a one and thus store information. But techniques for manufacturing the interior processes and linking them with the external world through conductors are problematic—electrical contacts are not easy to supply. Researchers do not yet know whether these artificial organic systems will stay cool or heat up when current flows through them. The rigid silicon crystal can handle quite a lot of heat, while organic compounds decay at comparatively low temperatures.

Solid-state research has provided increasingly exact methods for studying, measuring, and influencing atomic structures. In the future scientists will continue to apply these sophisticated methods to organic compounds and biological systems, forging a link between microelectronics and biotechnology. Forty years ago Erwin Schrödinger, one of the fathers of quantum physics, predicted in his now-famous book *What Is Life?* that understanding biological macromolecules, or "aperiodic crystals," would be imperative. Cracking the genetic code was the first major step in this direction.

Schrödinger called large organic molecules "aperiodic" because they lack the periodic atomic arrangement of inorganic materials such as silicon. The ever-recurring, orderly positioning of all the atoms on the regular lattice was a key to understanding solid crystalline materials. Understanding the much more complex and differentiated spatial arrangement of living matter is necessary before anything resembling a biochip can be constructed—a much more difficult task than understanding the periodic structure of the crystal.

Biotechnology and microelectronics approach each other from their own set limits. Both promise the adjoining of molecules to form building blocks for use in switching functions. The point of departure for microelectronics is the manipulable crystal grid, penetrating ever smaller dimensions, dividing and separating rather than adjoining. Microelectronics begins on a large scale and aims to reach the atomically small. Biotechnology has thus far made use of surprisingly few of the structuring methods of microphysics; instead, it has turned to the synthesizing art of organic chemistry. The increasing cooperation between these two disciplines in the coming decades will be interesting to watch.

The New Micromechanics

In addition to setting the pace for a possible future branch of biology, purified silicon has created the basis for a novel form of precision mechanics—micromechanics. Because silicon's symmetrical lattice can be constructed with the highest degree of precision, it has been used in minute tools. Nozzles with extremely fine openings can be created from single silicon crystals by etching away the unwanted portions with sophisticated microelectronics techniques. Thin membranes can be produced by etching silicon wafers, and these ultrafine membranes can serve as sensors for pressure measurements.

An entire instrument, for example, has been integrated into a silicon crystal lattice: a complete gas chromatograph for measuring the composition of gases has been put inside a silicon wafer. A conventional gas chromatograph is assembled from individual components connected by wires so gas can be transported electrically through pipes. In the new method the pipes become grooves etched in the silicon, with silicon or another material deposited on top. All necessary electrical functions are carried out with the technologies developed for integrated circuits. The micromechanics of silicon, with its tiny beams and reeds, its membranes and pores, can be linked directly with microelectronic functions.

Because semiconductors are highly sensitive to external influences—light, magnetic fields, foreign atoms, and mechanical stress—they make ideal indicators for gauges and probes. Silicon will most likely play a significant role in probes and sensors in which small oscillating membranes or mobile indicators and levers react to their environment, allowing very rapid measurements. Fine measurements of pressure are needed in, among other things, motor vehicles, washing machines, and environmental protection devices. Because of its versatility, silicon is used in 98 percent of all microelectronic devices today. Only in devices in which light plays a crucial role does silicon defer to other crystals.

The uses of silicon are expanding to new frontiers. It forms part of integrated devices so complex that computers are required to plan their construction. Giant software programs permit the planner to see the blueprint of a silicon circuit on his computer screen. The process of computer designing a new chip has become so precise that it obviates many experimental steps in the laboratory. The basic physical processes of silicon are so well understood that the interactions of its many structures can be simulated in the computer. Before any lab tests are conducted, scientists can try out a theoretical design on the computer and examine the likely electrical behavior of the final product.

Silicon has become both product and tool. Computers made of silicon use even more silicon in detectors to monitor the manufacture of integrated circuits. When the material is placed in the furnace, the heating process has to take place at just the right speed. The process of cooling the wafers down from more than 1,000 degrees centigrade—after oxidation, for example—also must be slow and steady. The shock of any sudden temperature change damages the crystal's perfect structure. A small motor, controlled by a microprocessor containing silicon, drives the silicon-laden quartz boat into or out of the furnace according to a precise schedule. Silicon chips in the computer's many other devices protect and guide the solid on its journey from raw crystal to finished chip. The role of the human worker, clothed in sterile white and avoiding the slightest dust, is just to stand by and watch.

Checking the finished products of today's integrated circuits naturally requires the help of integrated circuits. From the early days, when women workers using microscopes and instruments checked all the final numerical values, testing has developed into a fine art and its own specialized industry. Large computers, running through all the possibilities at rapid speeds, decide whether a chip will go into the rejects container. They even decide what laser surgery should be performed to rescue a chip with a birth defect.

On December 29, 1959, physicist Richard P. Feynman, a Nobel laureate, issued an "invitation to enter a new field of physics." The occasion was a meeting in Pasadena, California, of the American Physical Society. I remember it vividly, for it was the first meeting of this society that I attended. Feynman's address was titled, "There's Plenty of Room at the Bottom." As a particle theorist, Feynman thought in terms of very small dimensions, and he wanted to incite solid-state physicists to think much smaller. His challenge was, "Why cannot we write the entire twenty-four volumes of the *Encyclopaedia Britannica* on the head of a pin?" He treated this as a question of physical principles, for no fundamental law stood against such an achievement. Feynman then offered a prize of $1,000 to the first person who could reduce the information on a book page to fit an area $1/25{,}000$ smaller in linear scale so that it could be read by an electron microscope. He offered another prize for a $1/64$-cubic-inch electric motor. Feynman's talk is still enjoyable to read, even for the harried experimenter who has to carry out what he theorized. The challenges he set forth have been achieved, but there is still "plenty of room at the bottom" before we reach the point where physical principles forbid further miniaturization.

The Fifth Generation

Where is this competition leading us? Why house more and more memories in smaller and smaller spaces? The uses to which the increasingly powerful silicon chips can be put clearly make the technological advancements worthwhile. Tomography, for example, a method of examining the body without the use of a scalpel, would be impossible without high-performance computers. Even analyzing tomographical results requires powerful silicon chips. The hearing aid was the very first device to use the transistor, and now researchers feel it may soon be possible to link an amplifying sound sensor directly to the auditory nerves and guide this information to the brain. Microelectronics will also improve the working of artificial limbs. The technologist, using primitive means to process and transmit his images, can only admire the human eye, the most impressive processor of signals. A sharp, rapidly changing picture still contains too much information to be processed and transmitted digitally. Doing that will require larger and faster chips. If costs are sufficiently reduced, image transmission will generate a strong demand for silicon memory chips.

The Japanese have decided on an ambitious slogan for future silicon chip performance: "fifth-generation computers" are their long-term goal. The contours of their undertaking are not perfectly clear, but perhaps for that very reason the challenge is all the more demanding. According to the generation classification, we have developed four generations of computers in only forty years. The first generation still used those awkward electron tubes; the second, with its individually wired transistors, was already a solid-state device. The integrated circuit, equipped with just a few components, formed the third generation. According to the classification, VLSI (Very Large Scale Integration), is the fourth and present generation, with hundreds of thousands of individual elements on the crystal chip. At a 1981 conference in Tokyo the fifth generation was proclaimed as the next goal. If this new series lives up to expectations, it will probably test the limits of the crystal. The smallest dimensions, the least amount of heat generated, multilayered systems—perhaps a nearly three-dimensional filling of space—can be achieved only with new tools. This computer generation will be characterized not simply by the choice of components but by an entirely new way to process data. The fifth generation is expected to render obsolete the process of dissection into individual commands, as well as the need for user-unfriendly computer languages.

The Universal Challenge

Our world will find itself increasingly reflected in the microcosm of the crystal. Unlike previous methods of storing and transmitting information, the housing of data in the interior of a crystal is both intangible and inaccessible to our unaided senses. Microelectronics becomes more abstract as its scope enlarges.

Just as our civilization made a difficult but vital transition from bartering actual goods to using symbols for currency—from coins to credit cards—so will we now make the same transition in the exchange of information. Yet our adaptation to silicon's abstraction will be swift. A major part of human dealings already takes place solely in the exchange of electronic patterns. Data can be transformed into information in the tiniest crystallites on disks and tapes and then transported, stored, exchanged, and retransformed into the charge motion in the silicon chip. The increasing efficiency of this process will affect many aspects of our lives. Soon we will carry credit cards with built-in silicon chips. Indeed, microprocessors that change billing schedules have already altered the boundaries of the financial world.

Industry is undergoing upheavals as well. The construction company's head office can now send its factory directions and automated machine operating instructions on magnetic tapes. Bulky technical drawings and models are becoming a thing of the past. Office trays, stock lists for parts and tools, administrative files and other office fixtures are being replaced by computers. Card catalogs and voluminous reference works are quickly losing ground to electronic information retrieval. Today information can be surveyed and processed in more ways than ever before. Finer differentiations are possible. Precise insights and evaluations bring hidden connections to light. The Volkswagen Company is considering equipping its automobiles with a memory chip that will maintain a permanent list of all the car's structural specifications, maintenance requirements, and repairs.

Even more significant than the level of abstraction and coordination attained by microelectronics is the ability to create more complicated and flexible patterns of interaction. Today's traffic signals, for example, give clear but limited directions; those of the future, programmed by memory chips, will relay a wealth of information concerning traffic, weather, and accidents. A multinational research program in Europe, called PROMETHEUS, tackles these questions. Computer games are already impressive forerunners of what is to come, although their processes are often naive. The first chess computers were amazing, for they evidenced the machine's ability to interact with a human partner. Programming a computer to play chess was a true test of human skill.

We have always been fascinated by the concept of the labyrinth. Within us all lies the fear of becoming lost in an infinite tangle of alternatives. The maze represents the course of human destiny, shaped by the irreversible consequences of our decisions. The structure of the semiconducting memory chip, with its binary yes-no decision process, is also labyrinthine. The most intelligent computer games today take advantage of this mazelike character. A game's computer screen presents you with a fairytale world. Once your task is established, such as the rescue of a beautiful princess, you are given guidelines. You have to make decisions, enter new rooms and corridors, and use the objects placed at your disposal. In some cases you have to meet special logical requirements before taking the next step. Some of your mistakes cannot be rectified. The silicon chip and its magnetic memories remember the constellations of all these self-multiplying aspects. In this way the writer fulfills the classic dream of accompanying the reader into a maze of opportunities. Using only printed words, an author is incapable of such a feat; the reader's imagination must first overcome the linear rigidity of written words.

Today's computers have limits. For instance, it is still necessary to key an answer to a computer's question. And the computer's vocabulary, which is large but nonetheless limited, is a constraint. The images on the screen still lack high-quality definition. By contrast, fifth-generation computers are expected to respond to human speech and to provide moving pictures of the highest quality.

In describing the functions of a computer, Americans anthropomorphize it. Interaction between computers is called "handshaking"; chips are "personalized"; and the device has "intelligence." Computers that are easy to operate are called "user friendly." Americans see this terminology as part of the attempt to make the computer approximate the nearly attainable characteristics of the human being.

As computer functions more closely imitate human physical and intellectual attributes, the very word "computer" becomes misleading and inappropriate. The mere computation of numerical values is being overshadowed by new computer languages that permit the user to process, sort, and categorize names, properties, and relationships—not merely numbers. Progress toward man-machine communication is being facilitated not just by the growing speed of chip logic, but also by advances in the entire structure of the machine. New programming languages are pulling down the barriers between data and commands and paving the way for programs to generate new programs. Software is beginning to generate its own development.

Microelectronics presently accounts for just a few fractions of a percent of the gross national products of the big industrial nations. In the United States semiconductors are an $11.4 billion industry. One would scarcely notice that tiny sliver of the GNP pie, were it not for the extraordinarily high growth rates, coupled with falling costs. Microelectronics, omnipresent in business and industry, will become the key factor in commodities and labor markets.

The East-West conflict leaves its mark on microelectronics, which, in turn, influences the conflict. Gaining or maintaining the upper hand in armaments technology is unthinkable without microelectronics. The Soviet Union and its satellite states are conducting research and training, but they are finding it more and more difficult to obtain licenses. Many Eastern Bloc nations have had the bitter experience of obtaining an apparently cheap chip manufacturing license, only to find it totally useless two years later. Some nations have had to rely on goods smuggled from the United States or cheaply copied from American prototypes. According to many experts, rigidly structured economies are basically incapable of competing. Most Eastern Bloc research and development is done in isolated military laboratories.

The spread of microelectronics in the West will drastically widen the gap between the lifestyles and economies of the East and those of the West. Free access to data banks, unrestricted telecommunications, the influence of and on the media, personal computers and printers, satellite transmissions—all are part and parcel of Western life. But in countries where even a copier or telephone book is highly suspect and strictly monitored, such technologies are unheard of. Eastern European scientists worry that the gap in lifestyles will widen as the military powers of both sides gain even more influence.

One of my colleagues in the Max Planck Society had the opportunity in 1987 of conversing with some of Mikhail Gorbachev's technology advisers. They discussed the problem of modernization in such areas as personal computers, printers, and copiers. It was made clear that the reformers in the Soviet Union consider it very important to change the policy. "We want to distribute these tools as quickly as possible; with computers in private hands, there will be no way to turn back our reform plans." But many knowledgeable observers are doubtful that the reformers will be successful, because the entire Soviet political structure is at stake. At the 1987 American Physical Society meeting the physicist Yuri Orlov, who had recently emigrated, was invited to present his views. He distilled the questions down to a choice between introducing microelectronics into Soviet society, on the one hand, and maintaining control of the information flow, on the other. Orlov sadly predicted that the second choice was more likely, because giving up control of information is too high a price for the Soviets.

By mastering the atoms in the silicon crystal lattice, man has uncovered a promising microcosm—though one that is full of uncertainties. In the Middle Ages man saw himself as a microcosm that reflected the harmony of the macrocosm. As we approach the end of the twentieth century we no longer find "the heavens" a mysterious, sometimes frightening ruling power. Instead, we find ourselves part of a mechanistically determined cosmos. The atom, the new microcosm, now decides our definition of the world and our destiny in that world. We now look to plutonium nuclei, silicon, and other atoms for the secrets of creation and destruction, as well as harmony and symmetry. Now the macrocosm of society seeks the harmony of the microcosm of the atom. The quest for reconciliation of the parts with the whole has assumed a new character. We cannot look at the world with just one eye.

Bibliographical Note

. . . .

The following notes are intended for the reader who wants more detailed information about the various topics covered in this book.

1. CRYSTAL CRISES

The article "Ferdinand Braun: Forgotten Forefather" can be found in L. Marton and C. Marton, *Advances in Electronics and Electron Physics*, vol. 50 (New York: Academic Press, 1980), p. 241. A good biography is Friedrich Kurylo and Charles Susskind, *Ferdinand Braun* (Cambridge, Mass.: MIT Press, 1980). I will not try to list here the wealth of literature treating the electron tube, but I do want to mention Lee De Forest's autobiography, *Father of Radio* (Chicago: Wilcox and Follett, 1950).

Thomas S. Kuhn's influential book, *The Structure of Scientific Revolutions*, (Chicago: University of Chicago Press, 1962), is worthwhile reading. I. Bernard Cohen covers this topic in great detail in *Revolution in Science* (Cambridge, Mass.: Harvard University Press, 1985).

2. INNER SPACE

A comprehensive study of luminescence research is E. Newton Harvey's *A History of Luminescence* (Philadelphia: American Philosophical Society, 1957). I made a brief comparison of early and modern methods of investigating luminescence in the opening presentation of the International Conference on Luminescence held in Berlin in 1981; it is reprinted in *Journal of Luminescence* 24/25: 3.

Albert Einstein's Nobel Prize for physics in 1921 was not given for his theory of relativity, which at that time still seemed too controversial to the prize committee. The award was for his work on the photoemission of electrons from metals, with his concepts—radical for that era—concerning the quantum nature of light. Only after Bohr explained atomic structure did Einstein's light concepts gain general acceptance.

A good introduction to the history of modern physics is Emilio Segre, *From X-Rays to Quarks: Modern Physicists and Their Discoveries* (Berkeley: University of California Press, 1980). This book, however, discusses solids only in a paragraph on "Other Macroscopic Quantum Effects." The Göttingen School is honored in a symposium volume edited by Nevill Mott, *The Beginnings of Solid State Physics* (Cambridge: Cambridge University Press, 1980). The symposium, held in 1980 under the patronage of the British Royal Society, was organized by Mott, a Fellow of the Royal Society and a Nobel Prize laureate for physics. Ernest Braun contributed "The Contribution of the Göttingen School to Solid State Physics 1920–1940," and C. A. Hempstead reported on an interview with Pohl in 1974. Some of the remarks by Pohl quoted in this book were taken from Hempstead's chapter.

3. UNDER THE RADAR UMBRELLA

Sir Robert Alexander Watson Watt (1892–1973), the developer of radar technology for peacetime and defense uses, was an outstanding researcher. His books *Three Steps to Victory* (1957) and *The Pulse of Radar* (1959) describe how the new technology was used in the Battle of Britain. The British Royal Radar Establishment, still an important research facility, is taking on a larger role in coordinating microelectronics and communications technology in the United Kingdom. The books in the *M.I.T. Radiation Lab Series* present the radar research done in the United States during the war.

Two special issues of electronics journals provide in-depth, richly illustrated descriptions of the development of modern microelectronics. One is the special commemorative issue of *Electronics*, April 17, 1980. The professional journal of the International Institute of Electrical and Electronic Engineers, *IEEE Transactions on Electron Devices*, published a special issue in July 1976 on the history of electronics. In it William Shockley wrote about the invention of the transistor.

4. THE MOTHER OF INVENTION

The history of Bell Telephone Laboratories has been published in several volumes. Especially relevant for the development of semiconductors and the transistor is *Physical Sciences 1925–1980*, edited by S. Millman (AT&T Bell Laboratories, 1983).

Ernest Braun and Stuart MacDonald cover Bell Labs and the invention of the transistor in several chapters in *Revolution in Miniature* (Cambridge: Cambridge University Press, 1978). The two special issues of the electronics journals mentioned in the notes to Chapter 3 give particularly detailed historical descriptions.

Shockley's *Electrons and Holes in Semiconductors* (Princeton: Van Nostrand, 1953) was the first textbook in the new discipline of semiconductor physics. This classic work, which had a considerable impact on research and teaching, is still read and quoted today. International experts in the sciences are planning historical anthologies on a wealth of topics, including the problems of microelectronics. The International Project in the History of Solid State Physics is being coordinated by Lillian Hoddeson-Baym, Urbana, Illinois; Ernest Braun, Vienna, Austria; Spencer Weart, American Institute of Physics, New York; and Jürgen Teichmann, Munich.

5. TRANSISTORS COME OF AGE

Germanium researchers at Purdue University came close to making the same discoveries as those of Bell's much larger team. An interview with Purdue physicist Ralph Bray, who worked in semiconductor research, appears in Braun and Macdonald, *Revolution in Miniature*. In the final analysis, though, the little group headed by Professor Lark-Horovitz did not have a chance against the professional research at Bell Labs.

Alan Turing, the unconventional outsider, is portrayed by Andrew Hodges in *Alan Turing: The Enigma* (New York: Simon and Schuster, 1983). Critical readers will enjoy Tracy Kidder's book on the development of a new generation of computers, *The Soul of a New Machine* (Boston: Little, Brown, 1981).

6. SETTLING IN SILICON VALLEY

A recent, detailed account of this chapter in semiconductor history, *Silicon Valley Fever*, by Everett M. Rogers and Judith K. Larsen (New York: Basic Books, 1984). Technical journalist Dirk Hanson depicts the rise of Silicon Valley and its atmosphere in *The New Alchemists: Silicon Valley and the Microelectronics Revolution* (Boston: Little, Brown, 1982).

Soon after Fairchild was sold, a detailed history of the company appeared: "Fairchild Semiconductor: The Lily of the Valley," *Electronics News* 33 (September 28, 1987).

A little book that pokes fun at the style of language and work in the Valley is *The Official Silicon Valley Guy Handbook*, by Patty Bell and Doug Myrland (New York: Avon Books, 1983).

The Silicon Valley Genealogy poster, printed by Semiconductor Equipment and Materials Institute of Mountain View, California, shows the family tree of seventy-five firms that had arisen by 1981 from Shockley Transistor, the original business. Carolyn Caddes has published a collection of photographs, *Portraits of Success: Impressions of Silicon Valley Pioneers*, with a foreword by John Bardeen (Palo Alto, Calif.: Tioga Publishing, 1986).

Silicon is by far the most frequently discussed material in the scientific literature. In 1982 *Physics Abstracts* listed 3,300 publications giving new scientific results for this substance. Every three years the flood of literature seems to double.

7. COMPLETING THE CIRCUIT

Hanson, *New Alchemists*, as well as Braun and Macdonald's *Revolution in Miniature*, report on the dispute between Jack Kilby (Texas Instruments) and Robert Noyce (Fairchild/Intel), still unsettled, concerning patents and priority claims on the invention of the integrated circuit. The history of the development of the integrated circuit and the special contributions by Kilby and Noyce are well covered in T. R. Reid, *The Chip* (New York: Simon and Schuster, 1985).

The "law" devised by Gordon E. Moore describes the exponential development of integrated circuits, their continual doubling of density and halving of costs-per-transistor-function over particular time periods. Moore's Law does not provide a linear context, which would entail fixed cost differences for given time periods. Moore, who was one of the Shockley runaways who cofounded Fairchild, discusses these contexts in *IEEE Spectrum* 10 (April 1979): 30.

Very Large Scale Integration, edited by D. F. Barbe (Heidelberg: Springer-Verlag, 1980), provides a general yet expert survey of very large scale integration of semiconducting chips. In its last chapter, "VLSI in Other Countries," R. I. Scace outlines the situation in the "developing countries" of Europe. The classic textbook for semiconductors and chips is by Simon Sze, a Bell Labs employee: *Physics of Semiconductor Devices*, 2nd ed. (New York: John Wiley, 1982).

8. HANDOTAI SENSO

The Japanese use a wealth of catchwords to characterize their economic rivalry with the United States and the resulting tensions. *Handotai* (semiconductor) *senso* (war) is a particularly effective slogan, and the Japanese hesitate to use it in the presence of Westerners. Makoto Kikuchi's *Japanese Electronics—A Worm's Eye View of Its Evolution* (Tokyo: Simul Press, 1983), is insightful, surprisingly candid, and filled with personal anecdotes. The Japanese author counters his American colleagues' accusations against Nippon and gives his views of the principal differences between Japanese and American ways of thinking.

At irregular intervals American science and technology magazines, particularly *Scientific American*, publish twenty-four- or thirty-two-page advertising inserts titled "Japanese Technology Today." These intensive ad campaigns give high priority to electronics development and defend Japanese policies against the accusations of imitation, dumping, one-sidedness, and trade restrictions. The American book market has recently had a flood of publications on Japanese management strategies. Bruce Nussbaum's *The World after Oil: The Shifting Axis of Power and Wealth* (New York: Simon and Schuster, 1983), passes very harsh, controversial judgments on Japanese trade and research policies.

Japan promotes international conferences on future developments; there is, for example, a task force on future electronics devices. Publication of the proceedings of a meeting on superlattices, held in Tokyo in February 1984, was financed by the Keigin Society, which donated some of the profits from bicycle races and bets to this cause.

9. WHITHER AMERICA?

Reports on cooperative research ventures, such as the Semiconductor Research Corporation in North Carolina and the Microelectronics and Computer Technology Corporation of Austin, Texas, can often be found in newspapers and journals. See, for example, *Scientific American,* April 1986, p. 54, and *IEEE Spectrum* (April 1986): 75.

10. EUROPE ON THE SIDELINES

The estimates given in this chapter for per capita chip production were made by the Valvo Company for 1982. The import figures, from the Institut für Weltwirtschaft in Kiel, West Germany, appeared in *Zeit,* February 18, 1984. The newspaper *Frankfurter Allgemeine* published the comments of the Central Electrotechnical Industry Association on February 17, 1984. On February 4, 1984, the same paper published figures on the German-Japanese balance of trade.

An unsparing, sobering account of the state of microelectronics in Western Europe is provided by *Microelectronics in Western Europe: The Medium-Term Perspective 1983 to 1987,* edited by K. P. Friebe and A. Gerybadze (Berlin: Erich Schmidt-Verlag, 1984). A long and biting article on the finicky Germans appeared in the London *Financial Times* of February 8, 1983. An equally critical article about Europe from the Silicon Valley viewpoint was printed in the *San Jose Mercury* of February 12, 1984. In that year this theme became a central one in newspapers the world over.

Chapter 3 of Bruce Nussbaum's *World after Oil,* "The Decline of Germany and the Breakup of Europe" depicts Germany as a lost cause. In Nussbaum's opinion the nation heavily subsidizes smokestack industries and has no applied modern electronics industry. In 1976 a report by Mackintosh Consultants, entitled "Marktstudie Halbleiter" (Market Study: Semiconductors), revealed the sorry state of the European semiconductor industry and made numerous suggestions for catching up with the United States and Japan. The study was made for the German Federal Ministry of Research and Technology (NT-0682, December 1976).

11. THE CRYSTAL MICROCOSM

The von Klitzing effect is described in the April 1986 issue of *Scientific American* and in *Reviews of Modern Physics* 58 (1986): 519–531. The tunnel microscope, developed by Gerd Binnig and Heinrich Rohrer in IBM's Zurich laboratory, was first used to survey and measure atomic structures on surfaces. The young physicists who developed the microscope and made the findings won important international prizes, pointing up the competitive state of basic research in Europe.

The April 1983 issue of *Scientific American* examines the branch of microelectronics that employs exactingly processed silicon crystals. And finally, even today, years after the discovery of the double helix as the carrier of genetic information, Erwin Schrödinger's book *What Is Life?* (1943; reprint, Cambridge: Cambridge University Press), is worthwhile reading.

Index

. . . . . .